Contents

KT-440-443

Introduction

Welcome to Collins GCSE Science!

This Homework and Summary book is designed to help you get the most out of your GCSE Science studies. It covers GCSE Science, GCSE Additional Science, and GCSE Biology, Chemistry, Physics, all in one handy book!

There are several sections within the book. Here's how they work.

Summary content and Now Try This

This book is structured into topics which summarise the content you will learn on your course. The panels of summary points break all of the content down into short handy chunks to help you remember the key ideas. For more details, references to the relevant Collins Science textbook are also provided.

Within each summary panel there is a Now Try This box with practice questions. These questions are provided to help you check your own progress and understanding. Answers are provided at the back of the book.

At the top right of each page you will notice three circles. These are for you to mark in how well you feel you have understood everything on that page. For example:

 I have started this section but I don't completely understand or remember it all yet.

 I am getting better, but I have not yet been able to complete the Now Try This exercises correctly.

 I completely understand this section, and can complete all of the Now Try This exercises correctly.

Homework questions

Homework questions are provided for every topic. Your teacher will suggest which homework you should do.

Exam practice

The Exam-style questions pages will let you really prove you understand the Science, and will help you get ready for the exams. Note that these sample exam questions may cover content from more than one topic. This will help keep the key ideas you have learned fresh in your mind.

Model exam question answers

The Model answers are provided so you can see how questions should be answered.

We hope you find this Homework and Summary book useful. Good luck with your studies!

Your name _____

Class _____

Environment 1

CLASSIFICATION

- **Organisms** are **classified** by their **features**.
- **Variation** can occur within a **species** (intra-species) and between species (inter-species).
- A species is a group of organisms that share common features and are able to breed to produce fertile offspring.
- There are seven levels of classification: **kingdom**, phylum, class, order, family, genus and species.
- There are five kingdoms: plants, animals, bacteria, fungi and protista.

Now try this

a Fill in the missing words.

We classify organisms according to what they _____, when they are _____ and where they _____.

Differences between species are known as _____.

Oak trees are classified as plants because they produce their own _____.

 TOP TIP Use a mnemonic to help you remember the seven levels of classification.

Homework

1. List **five** characteristics that vary within the human species.
2. Name **two** characteristics for each of the **five** living kingdoms.
3. Donkeys and horses breed to produce infertile mules. Are mules considered a species? Which features do scientists use to classify new species?

FEEDING

- Feeding relationships within populations can be described using **food chains** and **webs**.
- The number of organisms at each stage of a food chain can be described using a **pyramid of numbers**.
- The mass of each stage can be described using a **pyramid of biomass**.
- Food chains always start with the **producer**.
- The mass decreases as you go up the food chain; this is because organisms at each stage use a portion of the **energy**.
- Energy is used for moving, maintaining body temperature and building up larger molecules. Energy is often lost as heat and movement.

Now try this

b Match the key words with the meanings.

i Consumer	Produces its own food by photosynthesis
ii Producer	The mass of all living matter in an area
iii Food web	The organism which is consumed by a predator
iv Biomass	Gets its energy by consuming other organisms
v Prey	Shows the flow of energy through a community

Homework

4. Draw a pyramid of numbers and a pyramid of biomass for the following food chain:

    oak tree → caterpillar → sparrow → fleas

5. Explain why both pyramids of biomass and number are useful for scientists.
6. Write an action plan for a government aiming to take its population out of famine. Is it better for farmers to grow wheat or raise chickens? Explain in terms of energy and biomass. Draw a food chain to illustrate your argument.

Edexcel GCSE 360Science

Homework and Summary Book

for Edexcel GCSE, Additional and Separate Sciences

Elizabeth Calton
Gurinder Chadha
Pam Large

William Collins' dream of knowledge for all began with the publication of his first book in 1819. A self-educated mill worker, he not only enriched millions of lives, but also founded a flourishing publishing house. Today, staying true to this spirit, Collins books are packed with inspiration, innovation and practical expertise. They place you at the centre of a world of possibility and give you exactly what you need to explore it.

Collins. Freedom to teach.

Published by Collins
An imprint of HarperCollins*Publishers*
77–85 Fulham Palace Road
Hammersmith
London
W6 8JB

Browse the complete Collins catalogue at
www.collinseducation.com

© HarperCollins*Publishers* Limited 2006

10 9 8 7 6 5 4 3 2 1

ISBN-13 978 0 00 721644 4
ISBN-10 0 00 721644 0

The authors assert their moral rights to be
identified as the authors of this work.

British Library Cataloguing in Publication Data
A Catalogue record for this publication is
available from the British Library

Commissioned by Cassandra Birmingham

Publishing Manager Michael Cotter

Project managed by Nicola Tidman

Edited by Anita Clark

Proofread by Lynn Watkins

Cover artwork by Bob Lea

Cover design by Starfish

Internal design and page make-up by JPD

Illustrations by JPD, Peter Harper, IFADesign
Ltd, Peters and Zabransky, Rory Walker, Pete
Smith and Peters Richardson (Beehive
Illustration) and Angela Knowles (Specs Art)

Production by Natasha Buckland

Printed and bound by the Bath Press, Glasgow
and Bath

Acknowledgements
The Publishers gratefully acknowledge the
following for permission to reproduce
photographs. Whilst every effort has been made
to trace the copyright holders, in cases where
this has been unsuccessful or if any have been
inadvertently overlooked, the Publishers will be
pleased to make the necessary arrangements at
the first opportunity.

p8 istockphoto; p9 Michael W. Tweedie /
Science Photo Library; p11 CNRI / Science
Photo Library; p12 Peter Menzel / Science
Photo Library; p18 Dr. John Brackenbury /
Science Photo Library; p21 istockphoto, CNRI /
Science Photo Library; p24 David Taylor /
Science Photo Library; p27 Charles D. Winters /
Science Photo Library; p29 James King-Holmes
/ Science Photo Library; p33 Paul Rapson /
Science Photo Library; p34 istockphoto;
p36 Robert Brook / Science Photo Library;
p37 Martin Bond / Science Photo Library;
p38 istockphoto; p43 istockphoto; p44 Jean-
Charles Cuillandre / Canada-France-Hawaii
Telescope / Science Photo Library;
p52 istockphoto; p57 © 2006 JupiterImages
Corporation; p59 Martin Bond / Science Photo
Library; p61 Andrew Lambert Photography /
Science Photo Library; p62 Erika Craddock /
Science Photo Library; p64 Maria Platt-Evans /
Science Photo Library; p69 Andrew Lambert
Photography / Science Photo Library;
p72 Andrew Lambert Photography / Science
Photo Library; p74 istockphoto, istockphoto;
p76 NASA / Science Photo Library;
p81 istockphoto; p86 istockphoto;
p87 istockphoto; p89 John Sanford / Science
Photo Library; p92 istockphoto;
p93 istockphoto; p95 istockphoto, istockphoto;
p96 Andrew McClenaghan / Science Photo
Library; p97 Andrew Lambert Photography /
Science Photo Library; Andrew Lambert
Photography / Science Photo Library;
p100 Andrew Lambert Photography / Science
Photo Library; p102 Andrew Lambert
Photography / Science Photo Library, Geoff
Tompkinson / Science Photo Library;
104 Adrienne Hart-Davis / Science Photo
Library; p106 L. Steinmark / Custom Medical
Stock Photo / Science Photo Library; p107 Hank
Morgan / Science Photo Library

ORGANIC vs CONVENTIONAL

- Conventional farming enables farmers to produce a **high yield** of crop but causes a great deal of damage to the **environment**.
- **Organic** farmers do not use chemicals on their crops. This type of farming produces a much lower yield of crop but is much more environmentally friendly.
- Chemicals build up as you go through the food chain. This is because the mass of organisms in each level of the food chain decreases, causing the chemical to get more and more concentrated. This process is called **bioaccumulation**.
- A pesticide is a chemical that kills pests; a **herbicide** is a chemical that kills unwanted plants.

Now try this

c Write **O** or **C** for organic or conventional.

Produces a higher yield of crop _____

Does not involve the use of pesticides and herbicides _____

Is more damaging to the environment _____

Crops are more expensive _____

Uses chemicals that are harmful to the environment _____

Uses natural fertilisers such as manure _____

Homework

7 Use a pyramid of biomass to explain why fatal amounts of toxins can build up in a food chain.

8 Another problem with using fertilisers is eutrophication. Find out what this means and explain its effect on the environment.

9 Organic farming enables us to protect our environment. Can this type of farming be sustained? Outline **two** arguments for and **two** arguments against organic farming.

COMPETITION

- There is **competition** within and between **species**.
- Organisms compete for light, space, mates, water and nutrition.
- Organisms within a community are often **interdependent**. This means that population size of one species may be affected by population size of another species within its food web.
- **Intraspecific competition** is competition within a species.
- **Interspecific competition** is competition between different species.

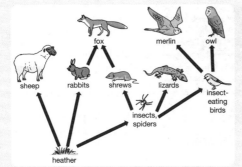

Now try this

d Use the food web on the left to help you answer the following.

If the number of rabbits decreases, the number of foxes _____.

If the number of foxes decreases, the number of shrews will _____.

If the number of owls decreases, the number of insect-eating birds will _____.

If the number of insects and spiders decrease, the number of shrews, lizards and insect-eating birds will _____.

Homework

10 List **four** factors that plants compete for.

11 List **four** factors that animals compete for.

12 Zebras and oxpeckers have a symbiotic relationship. Find out what this means. Describe what would happen to the oxpecker population if the zebras were to decrease in number.

Environment 2

A GROWING HUMAN POPULATION

- Scientists use **computer modelling** to predict what will happen to a **population**. This is particularly useful when trying to save a population from extinction.
- A number of factors affect population size, including disease, availability of space, light, water and food.
- Human population is growing **exponentially**. This is due to the increasing life expectancy caused by factors such as improved sanitation, healthcare and understanding of disease.

 TOP TIP The biggest cause of death in the UK 40 years ago was infectious disease; today the biggest cause of death is coronary heart disease.

Now try this

a Tick the factors that have caused an increase in the human population.

Birth rates ☐
AIDS ☐
Average life expectancy ☐
Tuberculosis ☐
Diet ☐
Improved sanitation ☐
Infectious disease ☐
Medicine ☐
Better understanding of disease ☐

Homework

1 Explain the difficulties faced when surveying populations in their habitat. How can computer modelling help us to overcome these difficulties?

2 How does a rapidly growing population affect the environment? How do large cities such as New York and London tackle the environmental effects of a growing population?

3 Explain why in the UK the number of people dying from infectious disease is rapidly declining but the number of people dying from coronary heart disease is rapidly increasing.

BECOMING EXTINCT

- **Evolution** enables organisms to **adapt** to better survive a changing **environment**.
- **Fossils** provide us with information about evolution.
- Fossil records give us clues as to how an organism has evolved.
- Organisms can become **extinct** if they are not well-adapted to their environment.

Now try this

b Fill in the missing words.

The evolution of organisms leads to new _____.

Older fossils are found in layers of sediment _____ younger ones.

Evolution is a _____ change in characteristics of a species.

If an organism is not well adapted to its environment it may become _____.

Homework

4 Explain how a growing human population threatens populations of other organisms.

5 Research how human activities led to the extinction of the dodo.

6 In 1994, extremely important fossils were discovered in China which provided the evolutionary link between birds and dinosaurs. Use the Internet to research these fossils and explain why their discovery was so important.

SURVIVAL OF THE FITTEST

- **Darwin's** theory of **natural selection** revolves around the fact that organisms are **adapted** to the environment in which they live.
- Organisms that are best adapted will be most successful and therefore will be most likely to **reproduce** and pass on their **genes**.
- Organisms that are poorly adapted will not survive and therefore their genes will be lost from the **population**.

 TOP TIP 'Survival of the fittest' ensures the organisms best adapted to their environment will be most likely to survive and pass on their genes.

Now try this

c Match the key words with the meanings.

i	Natural selection	Scientist who produced the theory of evolution
ii	Genes	Found in the nucleus of body cells, these determine your inherited characteristics
iii	Adaptation	Characteristic that enables an organism to better survive in its environment
iv	Evolution	Changing in order to become better adapted to an environment
v	Charles Darwin	Organisms best adapted to their environment will survive and pass on their genes

Homework

7   In the 19th Century, what evidence did Darwin have to base his theory on? What extra evidence do we have today that further proves his theory?

8   How has natural selection resulted in the camouflaging colours of animals such as the polar bear?

9   Why did Darwin's theory of evolution attract such controversy when it was first published?

GENETIC MODIFICATION

- For thousands of years humans have been carrying out **selective breeding**.
- Selective breeding is when animals or plants are chosen for breeding according to their **characteristics**. This produces **offspring** with features that benefit us.
- **Natural selection** and selective breeding are similar. In natural selection, positive characteristics are chosen by their suitability to a certain environment; in selective breeding, characteristics are chosen by humans.
- **Genetic modification** is a much quicker process. **Genes** coding for desired characteristics are identified and inserted into gametes of an organism.

TOP TIP In a sense we have been genetically modifying for thousands of years; selective breeding involves choosing plants and animals for their characteristics.

Now try this

d Label each of the following statements **natural (N)**, **selective (S)** or **genetic (G)**.

Environment selects the genes _____

Humans breeding organisms for characteristics advantageous to us _____

The fastest type of selection _____

This type of selection has resulted in the evolution of species _____

Identifying genes and inserting them directly into gametes _____

Involves human intervention _____

Homework

10   Which features might a plant breeder select for when breeding a crop such as corn?

11   Why is selective breeding in plants a much faster process than in animals?

12   Selective breeding, natural selection and genetic engineering have many similarities. Compare and contrast these processes.

Genes 1

GENES

- Genetic information is found in the **nucleus** of all cells.
- Human cells have 23 pairs of **chromosomes**.
- Chromosomes are made up of **DNA**.
- **Genes** carry information that determines all **inherited characteristics**.
- A gene is a section of chromosome which 'codes for' a certain characteristic.

 TOP TIP Chromosomes give the cell instructions on how to make certain proteins.

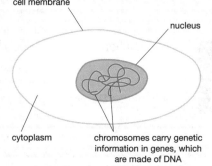

cell membrane

nucleus

cytoplasm

chromosomes carry genetic information in genes, which are made of DNA

Now try this

a Fill in the missing words.

Twenty-three pairs of _____ are found within the nucleus of the cell. These help control cell activity. _____ are sections of a chromosome which code for a certain _____. Chromosomes are made up of a chemical called _____. _____ characteristics are passed from parent to offspring via the inheritance of certain genes.

Homework

1 Make a list of inherited human characteristics.
2 Define the following words: chromosome, gene, DNA, allele.
3 Find out what DNA stands for. Who discovered it and when? Why was its discovery so important?

SEXUAL AND ASEXUAL

- Organisms **reproduce** by either **sexual** or **asexual** reproduction.
- Asexual reproduction involves only one parent. The offspring are **clones** of the parent.
- Sexual reproduction involves two parents.
- **Gametes** (sex cells) contain half the number of **chromosomes** of a normal body cell.
- The production of gametes involves a type of **cell division** called **meiosis** – this creates cells with only half the normal number of chromosomes.
- **Fertilisation** results in the production of new offspring by sexual reproduction; it involves the fusing of a male and a female gamete.

Now try this

b Label the following **sexual (S)** or **asexual (A)**.

Offspring are genetically identical to parents	_____
Involves gametes	_____
Produces very little variation	_____
Involves only one parent	_____
Offspring are not genetically identical to parents	_____
Clones are produced	_____
Fertilisation occurs	_____

Homework

4 Explain the difference between asexual and sexual reproduction. Give **one** example of an organism that reproduces asexually and **one** that reproduces sexually.
5 Why do gametes contain only half the genetic information that a normal body cell does?
6 The spider plant (*Chlorophytum*) reproduces asexually. Why are its offspring clones of the parent?

INHERITANCE

- There are many versions of **genes** – these are called **alleles**.
- Every characteristic is coded for by two alleles (one from the mother and one from the father). These may be two different alleles or two the same.
- Some alleles are **dominant**. If a dominant allele is present it will always show up.
- Some alleles are **recessive**. This characteristic will only show up if both alleles are present.
- If a **genotype** contains two identical alleles, we call this **homozygous**.
- If a genotype contains two different alleles, this is **heterozygous**.

 Each characteristic is coded for by two genes. Each version of a gene is called an allele.

Now try this

c Fill in the following Punnett square for eye colour.

	B	b
B		
b		

What is the chance of these parents having a blue-eyed child? _____

What is the chance of these parents having a brown-eyed child? _____

What colour are the mother's eyes? _____

What colour are the father's eyes? _____

Homework

7 Research cystic fibrosis. What are the symptoms?

8 What do 'phenotype' and 'genotype' mean?

9 Cystic fibrosis is a recessive inherited disease (c). Calculate the probability of a child being born with cystic fibrosis to parents who are both carriers (heterozygous Cc).

MUTATIONS

- A **mutation** has occurred when a **gene** has changed and no longer codes for its **characteristic** correctly.
- Some mutations can result in **inherited diseases**.
- People can either be **carriers** or **sufferers** of a disease.
- Carriers may not be aware of their genetic mutation.
- **Haemophilia** is a genetic disease where a mutation on the **X chromosome** results in the blood being unable to clot properly.
- **Cystic fibrosis** is a disease that affects many organs in the body, clogging them with a sticky mucus.

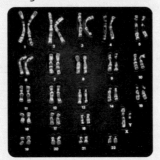

Now try this

d Fill in the missing words.

Genes code for certain _____.
If the structure of the gene is changed, we call this a _____.
_____ is a genetic disorder resulting in the blood being unable to clot properly. _____
_____ is a disease that affects many organs in the body. A sticky mucus is produced.

 Females have two X chromosomes; males have an X and a Y chromosome.

Homework

10 The allele for blue eyes is recessive and the allele for brown eyes is dominant. For a person to be blue-eyed, what genotype must they have?

11 Why can females carry haemophilia but not suffer from it themselves?

12 Sickle cell anaemia is a mutation affecting the red blood cells. Find out which parts of the world have populations that suffer from this disease. Explain why this mutation can be an advantage to certain populations.

Genes 2

GENES AND THE ENVIRONMENT

- **Variation** between organisms of the same species can be **inherited**. This is where certain traits are passed on from parent to offspring.
- Some variation is not passed on from parent to offspring – this is known as **environmental variation**, for example the length of your hair.
- Some inherited traits are also affected by the environment. Human height is an inherited trait but poor diet or disease can lead to stunted growth.

 Some variation is caused by a combination of environmental and inherited factors.

Now try this

a Are the following **inherited (I)** and/or **environmental (E)**?

Height	_____
Tattoos	_____
Weight	_____
Dress sense	_____
Pierced ears	_____
Eye colour	_____

Homework

1 How are plants affected by their environmental conditions?

2 Explain how, although height is an inherited characteristic, it can also be affected by environment.

GENE THERAPY

- **Genetic engineering** has enabled us to create **genetically modified organisms (GMO)**.
- We have been able to create cows whose milk contains antibodies that are useful for humans and cows that produce healthier, low cholesterol milk.
- We have also been able to make pigs that produce insulin to treat diabetes.
- Scientists identify the **gene** they want and cut it out using **enzymes**. It is then placed into the nucleus of the cells of the new organism. **Clones** are then made from this new organism.

① Isolate the relevant gene in a DNA molecule
② Cut out the gene using restriction enzyme
③ Produce millions of copies of the gene
④ Insert the gene into a carrier - perhaps a virus or a liposome
⑤ Spray the gene-carrying particles into nose of a sufferer
⑥ The particles cross into cells in the lungs carrying the working gene with them

Now try this

b Put the following steps in order.

- ☐ GMO is born displaying the characteristic of its new gene.
- ☐ The gene is carefully inserted into the nucleus of the fertilised egg before it undergoes its first division.
- ☐ The egg develops into an embryo.
- ☐ Identify the required gene.
- ☐ The embryo is inserted into the surrogate mother.
- ☐ Use special enzymes to cut the gene out.

Homework

3 Write a paragraph explaining the advantages for GMO.

4 Write a paragraph describing the arguments against GMO.

5 Some people disagree with GMO but this may be because they do not understand what it is. Prepare a presentation giving the facts about GMO.

THE HUMAN GENOME PROJECT

- The **human genome organisation (HUGO)** was created in 1989. Its aim is to identify and study all the **genes** in the **human genotype**.
- This has many positive implications, such as early diagnosis of cancer and other genetic disorders.
- **DNA** has been particularly useful in **forensic science**; every person has their own unique genetic code.
- This also has a number of potentially negative implications such as parents being able to decide what their baby will look like.
- With careful regulation, the human genome project could benefit human society. Many people, however, worry that it may not be regulated carefully, creating a great number of problems.

Now try this

c Fill in the missing words.

The human genome project has been identifying the position of all the _____ found in the human genome. This means that genetic diseases could be _____. Genetic disorders such as _____ _____ and _____ will become things of the past. Some people feel that genetic engineering is unethical, especially if we start to change the genes that control our _____ or _____.

Homework

6 Name **two** of the positive implications of the human genome project.

7 Name **two** of the negative implications of the human genome project.

8 Do you think people should be able to choose characteristics of their unborn offspring? Explain your answer.

THE ETHICS OF GENETIC MODIFICATION

- **Genetic engineering** could help humans:
  - grow better crops and farm animals
  - cure genetic diseases
  - produce useful substances (for example, insulin).
- Transplant patients must take drugs for the rest of their lives to stop organ rejection. The technology to clone close-match tissue is becoming more of a reality.
- Geneticists are developing **cures** for breast cancer and cystic fibrosis through **gene therapy**.
- There is much ethical debate about genetic modification being used by parents to predetermine the traits of their children.
- People opposing genetic modification worry that use of this method for curing genetic disorders may lead to it being used for altering genetic features, such as intelligence.

Now try this

d Circle the genes that you believe we should be allowed to alter.

Hair colour   Dwarfism

Height   Eye colour

Intelligence   Albino

Cystic fibrosis   Skin colour

Downs Syndrome   Cancer

Do your results match your neighbour's?

Homework

9   Do you believe it is acceptable to clone an animal to create body parts?

10   Why are the public worried about the consequences of being able to genetically test IVF embryos before they are implanted into the womb?

11   How is gene therapy providing a treatment for cystic fibrosis?

Electrical and chemical symbols 1

THE NERVOUS SYSTEM

- The **central nervous system** comprises of three types of **neurone**:
  - **sensory**: sense and send messages to the spine
  - **relay**: connect the sensory and motor neurones
  - **motor**: carry the message which elicits a response to the **stimulus**.
- Parkinson's disease is a disorder where important neurones in the brain are damaged or die. Sufferers become less able to control their muscles.
- Overactive neurones can result in epilepsy.
- A stroke occurs when the blood supply is cut off to an area of the brain, resulting in neurones being damaged or dying.
- We can measure **reaction time** by dropping a ruler between a person's fingers and measuring the distance it takes for them to catch the ruler.

Now try this

a Tick the factors that must be kept constant when investigating reaction time.

Alcohol intake ☐
Length of ruler ☐
Caffeine intake ☐
Type of ruler ☐
Age of subject ☐
Sex of subject ☐
Temperature of room ☐
Height of subject ☐
Intelligence of subject ☐

Homework

1 Draw a diagram showing the structure of a motor nerve.

2 Draw a diagram showing the structure of a sensory nerve. How does this differ to a motor nerve?

3 Design an experiment to measure the effect of caffeine on reaction times. What problems might you encounter and how could they be overcome?

REFLEX ACTIONS

- A **reflex** is an involuntary **response** to a certain **stimulus**.
- A **reflex arc** shows the pathway from when the body receives a stimulus resulting in a reflex.
- Some reflexes are learned, for example your responses when driving a car.
- Other reflexes, such as ducking when something travels close to your head, are automatic. Such responses play an important part in survival.
- A **receptor** senses a stimulus, for example the skin senses the temperature outside the body.

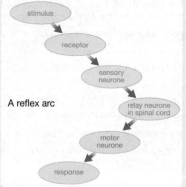

A reflex arc

stimulus → receptor → sensory neurone → relay neurone in spinal cord → motor neurone → response

Now try this

b Fill in the missing words.

When it is cold, the body responds by _____ and _____ stand up on end. When it is too dark, the iris responds by _____ in size. Many reflexes, such as _____ when something comes near to your head, play an important role in _____.

Homework

4 Draw the reflex arc showing what happens if you accidentally place your hand on a hot stove.

5 Explain the difference between a voluntary and a reflex action. Give **one** example of each.

6 How might you investigate the speed at which nerves carry messages? (Hint: you may need to use more than one person.)

See pages 62-71 of Collins GCSE Science

HOMEOSTASIS

- Many internal conditions of the body must be kept in a state of **homeostasis** (kept constant). For example, temperature, pH, carbon dioxide and glucose concentrations.
- Temperature is controlled by an internal sensor called the **hypothalamus**, in the brain.
- If the blood flowing through the hypothalamus is too warm, the body responds by trying to cool down.
- If the blood flowing through the brain contains a high concentration of carbon dioxide, the body responds by increasing the ventilation and heart rates.

Now try this

c Match the key words with the meanings.

i Receptors	Keeping the internal conditions of the body constant
ii Blood	Example: light
iii External stimulus	Example: blood glucose levels
iv Hypothalamus	Centre in the brain which monitors body temperature
v Internal stimulus	Carries hormones around the body
vi Homeostasis	Found in the skin

Homework

7 How does the body respond to the core body temperature being too high?

8 How do the eyes respond to a low level of light?

9 Which external and which internal factors does the body sense?

BLOOD

- **Blood** is made up of **plasma**, **white blood cells**, **red blood cells** and **platelets**.
- Plasma is a straw-coloured liquid.
- Platelets are involved in **blood clotting**.
- White blood cells produce **antibodies** and form part of the **immune system**.
- Red blood cells are biconcave-shaped and contain **haemoglobin**. They carry **oxygen** around the body.
- There are three types of blood vessel – **arteries**, **veins** and **capillaries**.

Arteries carry blood **A**way from the heart. Ve**IN**s carry blood **IN**to the heart.

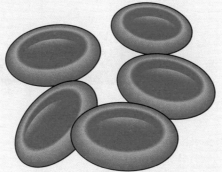

Now try this

d Fill in the missing words.

The function of the blood is to carry away waste such as _____ _____ and _____, take _____ to the body cells for respiration, and carry _____ such as insulin around the body to target organs.

Homework

10 Draw and describe white and red blood cells.

11 Draw a cross-section of an artery and a vein.

12 Explain why you cannot feel a pulse in veins but you can in arteries.

Electrical and chemical symbols 2

HORMONES

- **Hormones** are chemicals that are carried around in the blood and affect specific **target organs**.
- Hormones such as **adrenalin** are released as a **response** to an **external stimulus**. Hormones such as **oestrogen** are part of the normal routine of the body.
- Examples of hormones include oestrogen, testosterone, insulin and adrenalin.

Now try this

a Label the following **reflex action (R)** or **hormone (H)** action.

Puberty	_____
Responding to changes in the levels of light	_____
The menstrual cycle	_____
Moving your hand away from a hot pan	_____
Coughing	_____

Homework

1 How do hormones affect the body during puberty?
2 Which processes are controlled by hormones?
3 Find out about the major glands in the endocrine system.

HORMONES AND FERTILITY

- The **menstrual cycle** is a 28-day cycle controlled by **hormones**.
- **FSH** causes the egg to mature.
- **LH** stimulates **ovulation**.
- Rising levels of **oestrogen** inhibit the production of FSH and increase the production of LH.
- **Progesterone** maintains the uterus wall and prevents the further release of eggs.
- We use hormones to control **fertility**. FSH is given to women to increase fertility and stimulate the production of eggs.
- Oestrogen is given as a **contraceptive** – it reduces the production of FSH so eggs do not mature.

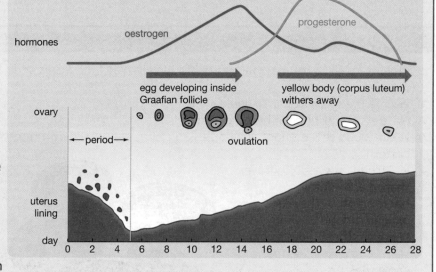

Homework

4 When in the menstrual cycle are oestrogen levels highest?
5 What is the role of progesterone in the menstrual cycle and during pregnancy?
6 Which hormone is given to women during IVF treatment to encourage their bodies to produce more eggs?

IVF

- **IVF** is an expensive fertility treatment.
- First, the mother undergoes **hormone treatment** so that she produces many **eggs**.
- Eggs are then removed and **fertilised** in the laboratory.
- Fertilised eggs are **implanted** back into the mother and allowed to develop.
- IVF is not 100 per cent successful and many couples have to go through numerous cycles of treatment before having a baby.

Now try this

b Fill in the missing words.

IVF is a process by which _____ occurs outside the body. The mother is given _____ treatment to encourage her to produce more _____ cells. These are removed and combined with _____ cells from the father. The fertilised egg divides and is replaced into the mother's _____.

Homework

7 What does IVF stand for? What does this mean?

8 Should IVF be offered on the NHS?

9 Write **four** arguments for and **four** arguments against the view that post-menopausal women should not be given IVF treatment.

CONTROLLING DIABETES

- **Insulin** is a **hormone** produced by the **pancreas**. It regulates the **blood glucose** levels.
- If the blood glucose is too high, insulin is produced. This causes glucose to be broken down.
- If insufficient insulin is produced or if insulin receptors do not function properly (often as a result of a diet high in sugar), the blood glucose levels can become too high.
- **Diabetes** is treated by a combination of diet and insulin injections.
- Initially insulin was extracted from fish, pig or calf pancreases but this sometimes caused an allergic reaction.
- Today, human insulin is made by genetically modified bacteria.

Now try this

c Label the following diagram to explain how the insulin gene is inserted into a bacterium.

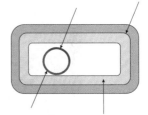

Homework

10 What is the simple test that doctors carry out to determine if someone is diabetic?

11 Research the symptoms of diabetes.

12 Insulin was initially extracted from fish, calf or pig pancreases. How has the introduction of genetically modified bacteria helped the treatment of diabetes?

Use, misuse and abuse 1

MICROBES AND DISEASE

- A **microorganism** is any organism too small to be seen with the naked eye.
- There are three types of microorganism: **bacteria**, **fungi** and **virus**.
- Bacteria do not have a nucleus; they are very simple cells. **Infections** caused by bacteria can be treated with **antibiotics**.
- Viruses are not cells. They are a **protein coat** containing genetic material. Viruses infect body cells with their genetic information.
- Fungi are spread by **spores**. They grow in moist, dark environments. Skin diseases such as athlete's foot and ringworm are examples of diseases caused by fungi.

 A **pathogen** is any organism that causes a disease.

Now try this

a Are the following **bacteria (B)**, **fungi (F)** or **virus (V)**?

Has a cell wall	_____
Flu	_____
Reproduces by creating spores	_____
Not many features of a cell	_____
Athlete's foot	_____
Can be treated using antibiotics	_____
Ringworm	_____
HIV	_____
Grow in dark, damp conditions	_____
Can be useful as well as disease-causing	_____

Homework

1 Draw a diagram of a virus, bacterium and fungus. Is a bacterium an animal or a plant? Explain your answer.

2 Research why viruses cannot be treated with antibiotics.

SPREADING DISEASE

- **Disease** can be spread by direct and indirect contact.
- **Pathogens** can be:
  - airborne – for example, transmitted through a sneeze
  - water-borne – for example, cholera is passed through dirty water
  - food-borne – often in undercooked food
  - **vector**-borne – passed via another organism, for example malaria.
- Pathogens can be passed from mother to child through the placenta or through breast milk. They can also be sexually transmitted.

Now try this

b Fill in the missing words.

_____ can enter through openings in the body such as the _____, _____ and _____. If enough microbes enter, this can result in a _____. Pathogens can be transmitted indirectly (via a _____) or through _____ contact. Pathogens can be transmitted through the _____, _____, or water.

Homework

3 Name **one** disease transmitted via each of the following: air, food, water and vector.

4 Schistosomiasis (also known as bilharzia) is a disease caused by a worm. Find out how this disease is transmitted?

See pages 84-91 of Collins GCSE Science

KEEPING MICROBES OUT

- The body has three lines of defence in order to prevent **pathogens** from entering the body.
- The first line of defence includes the skin, hairs in the nose and cilia in the trachea. This also includes chemical barriers such as enzymes in tears, which prevent pathogens from entering the eyes.
- If a pathogen does manage to enter the body, a non-specific second line of defence takes its effect.
- **White blood cells** recognise bacteria as being foreign bodies and **ingest** them in order to prevent infection from setting in.
- Inflammation containing dead bacteria and pus is likely to occur.

 If a large number of microbes enter the body, this can result in disease.

Now try this

c Match the key words with the meanings.

i	Engulf	An enzyme found in tears
ii	Cilia	Any organism not visible to the naked eye
iii	Lysozyme	Small hairs found in the trachea to prevent microbes from entering the lungs
iv	Microbe	Cell with a very large nucleus
v	White blood cell	Any disease-causing organism
vi	Pathogen	White blood cells _____ bacteria

Homework

5 Label a diagram of the body showing the first line of defence.

6 How do bacteria found in the vagina prevent yeast infections?

7 If you cut yourself, the area often becomes inflamed – explain why.

IMMUNITY

- **Antigens** are specific **proteins** found on the surface of all cells.
- **White blood cells** recognise foreign antigens. When this happens, they produce **antibodies** specific to the antigen.
- These antibodies stick to the antigen causing the pathogen to clump together. This provides a signal to other white blood cells to engulf and ingest the pathogen.
- If an antigen has been met before, white blood cells are able to produce specific antibodies more quickly.
- Some **vaccines** work by exposing the **immune system** to an inert version or part of a pathogen. This enables the body to respond more quickly when it does meet the pathogen.

Now try this

d Match the key words with the meanings.

i	Engulf	Proteins found on the surface of foreign cells
ii	Antibody	Released by white blood cells as a response to a pathogen
iii	White blood cell	These cells survive much longer than other white blood cells, and are able to elicit a fast response to a pathogen that has been met before
iv	Memory cell	The ability to resist a certain disease
v	Antigen	White blood cells _____ and ingest bacteria
vi	Immune	Cells found in the blood which play a large part in immunity

Homework

8 Explain how your body can acquire immunity to disease.

9 What is a vaccination? How do they work?

10 Research HIV. Why is the body unable to elicit an immune response to this virus?

Use, misuse and abuse 2

PREVENTING DISEASE

- **Immunity** is the ability to resist disease.
- Immunity can be acquired in one of three ways:
  - via breast milk
  - through **vaccination**
  - through being exposed to the disease.
- Breast milk contains vital **antibodies** which assist newborn babies in the development of their **immune system**.
- Vaccinations work by giving a dose of a dead or synthetic version of the **pathogen**. This enables the body to have an immune response and be prepared should it meet this pathogen in future.
- Once the body has been exposed to a pathogen, special cells called **memory cells** retain the ability to produce specific **antibodies** for it. This enables the body to respond quickly should it meet this pathogen again.

Now try this

a Fill in the missing words.

A _____ is anything that causes disease. Immunity is the ability to resist _____. A vaccine provides the body with artificial immunity by exposing the _____ system to a deadened or synthetic version of the _____. This enables the body to produce the correct _____ quickly when it does meet the pathogen.

Homework

1 Why is it advantageous for mothers to breastfeed their newborn babies?
2 Write the story of Edward Jenner's discovery.
3 Why is it unlikely that somebody will suffer from chicken pox twice?

TB

- **Tuberculosis** (TB) is a disease caused by a **bacterial infection** of the lungs. This infection can spread to other parts of the body and can be fatal. It was once a major killer.
- TB is **airborne** and is spread through sneezes and coughing.
- TB leaves the sufferer with terrible scarring to the lungs. It can be diagnosed by X-ray.
- Since the 1940s, the number of cases of TB has been gradually dropping due to the introduction of **antibiotics** to treat the disease and the BCG **vaccination**.
- More recently, cases of TB have been gradually rising due to:
  - the evolution of drug-resistant bacteria
  - the increase in HIV sufferers
  - overcrowded living conditions in cities
  - the price of medication.

Now try this

b Circle the symptoms of TB.

Vomiting

Coughing up blood

Fatigue

Night sweats

Headache

Fever

Rash

Scarring of the lungs

Weight loss

Cancer

High blood pressure

Homework

4 How was the development of antibiotics vital in the fight against TB?
5 Why is TB particularly prevalent in the homeless and in large cities?
6 Find out how the emergence of TB has been affected by the increase in HIV sufferers.

DRUGS

- A **drug** is a substance that elicits a change in the body.
- Some drugs are physically **addictive**, meaning the body comes to rely on their usage for normal function. Drugs may also be psychologically addictive.
- Some drugs promote **nerve signals** across the **synapse**, for example **caffeine**. Such drugs make you feel more alert.
- Other drugs block signals from crossing the synapse, for example **heroin**. This results in the blocking of pain.

Now try this

c Are the following statements **true** or **false**?

All drugs are physically addictive _____

Caffeine is an example of a drug _____

Paracetamol is a painkiller; it promotes signals to be sent across nerve synapses _____

Alcohol increases your reaction rate _____

Nicotine is highly addictive _____

Homework

7 Why can solvents be so dangerous? Explain their effect upon the body.

8 Drugs such as opiates are used for pain relief for terminally ill patients. Do you think this is acceptable? Take into account the fact that these drugs are particularly addictive.

9 Draw a diagram of a synapse. Use this to explain how stimulants, such as caffeine, and painkillers, such as paracetamol, work.

SMOKING AND ALCOHOL

- **Alcohol** and **tobacco** are both **legal drugs**.
- Alcohol is a drug made through the **fermentation** of glucose.
- Alcohol is a **depressant** (it blocks signals across the synapse).
- Excessive alcohol abuse can lead to liver damage. This organ is involved in removing dangerous toxins from the blood.
- **Nicotine** found in cigarettes is highly **addictive**.
- Cigarettes contain thousands of chemicals. Many of these are **carcinogens**, meaning they cause cancer.

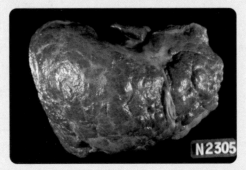

N2305

Now try this

d Fill in the missing words.

A _____ is a substance that has a chemical effect on the body. Alcohol is a _____. It is a _____ (meaning it _____ signals across the synapse). If you drink alcohol, your _____ time increases. The long-term effects of alcohol include cirrhosis of the _____.

Homework

10 What do we mean by 'binge drinking'?

11 Make a list of chemicals found in cigarettes. What does 'cancerous' mean?

12 Alcohol abuse can lead to cirrhosis of the liver. What is the function of this organ?

All about atoms

INSIDE ATOMS

- Every substance is made from **atoms**.
- Inside an atom, negative **electrons** fly around a tiny **nucleus**.
- The nucleus contains positive **protons** and neutral **neutrons**.
- **Atomic numbers** show how many protons there are.
- The number of protons decides the **element**.
- In a neutral atom, electrons are equal to protons.
- The three **shells** nearest the nucleus hold 2, 8 and 8 electrons.

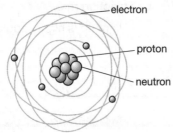

electron

proton

neutron

Now try this

a Match the statements on the right to the particles on the left.

Proton	Negative
	Around the outside of an atom
	Positive
Neutron	In the nucleus
	Neutral
	Number different for every element
Electron	

 In the periodic table, elements are in order of **atomic number**.

Homework

1 Prepare a presentation to show what atoms are like.
2 Search for some record breakers, for example atoms in the most expensive element.
3 Who discovered that atoms have nuclei? How was this discovery made?

SYMBOLS AND FORMULAE

- Every **element** has a **symbol** in the **periodic table**.
- If symbols have two letters, only the first is a capital.
- Elements combine chemically to make **compounds**.
- Compound names usually end in –ide.
- Compounds with two elements plus oxygen often end in –ate.
- When compounds have more than one of each atom, little numbers are used, for example CO_2 has one carbon and two oxygen atoms.

 Each element starts with a capital. **Co** is one atom of an element, but **CO** is two atoms in a compound.

Now try this

b Circle the correct answer.

An element:
H_2O CO_2 CO Co

A compound:
Ca Ni NO He

Formula with three atoms:
NH_3 CaO NaCl H_2O

Could end in –ate:
CO_2 CuO $CuSO_4$ Na_2O

Ends in –ide:
S CuS H_2SO_4 $FeSO_4$

Homework

4 What does the formula tell you about: NO_2, Zn, $CaSO_4$, Na_2S, HNO_3?
5 List some names ending in –ide and –ate that are found on household labels.
6 Find out why some symbols are not the first letters of the element's name.

WRITING EQUATIONS

- A **reactant's** atoms get rearranged when **products** form.
- The number of atoms does not change, nor does the mass.
- The same atoms are present before and after the arrow.
- Numbers show there is more than one copy of a formula.
- H_2 means H–H and $2H_2$ means H–H + H–H (two molecules, four atoms).

Now try this

c Complete or balance the following equations.

iron + oxygen → _____

copper + sulphur → _____

potassium + chlorine → _____

$H_2 + Br_2 \rightarrow$ ___HBr

___$Mg + O_2 \rightarrow$ ___MgO

___$H_2 + O_2 \rightarrow$ ___H_2O

(not to scale)

H_2 + Cl_2 $\longrightarrow$ **2HCl**

TOP TIP You can only put numbers **in front** of formulae when you **balance** an equation.

Homework

7 Design a homework sheet on writing equations.

8 Prepare a presentation to show how to balance equations.

9 Find some reactions that explode and write the equations.

RELEASING ENERGY

- Some **reactions** are faster than others.
- **Reaction rates** increase as the temperature rises.
- Every substance contains **chemical energy**.
- **Endothermic** reactions, like cooking, take in heat. The products have more energy than the reactants.
- **Exothermic** reactions, like burning, give out heat. The products have less energy than the reactants, but energy may be needed to start the reaction.

Now try this

d Write **exothermic** or **endothermic** for each reaction.

Reactants	Temperature reached (°C)	Type of reaction
methane + oxygen	350	
sodium + water	90	
acid + bicarbonate	–6	
magnesium + acid	70	

Homework

10 Draw a design for a hand warmer and explain how it works.

11 What properties do the reactions that take place in fireworks need to have?

12 Research some of the reactions that take place in cold packs and hand warmers.

The elements

DETECTING METALS

- Sodium hydroxide is an **alkali**.
- Alkalis form **precipitates** with **metal** compounds.
- Precipitates are solids formed when **solutions** mix.
- **Transition metals** make coloured precipitates: Zn (white), Cu (blue), Fe(II) (green), Fe(III) (brown).
- **Flame tests** can identify the metals in some compounds: Na (yellow), K (lilac), Li (crimson), Cu (green), Ca (orange).

a Match each result to the metal in the compound.

i	Blue precipitate	Cu
ii	Yellow flame	Fe(II)
iii	Green flame	Li
iv	Green precipitate	Fe(III)
v	Crimson flame	Na
vi	Brown precipitate	
vii	Orange flame	Ca
viii	Lilac flame	K

 TOP TIP The tests used to identify metals only work when the metal is part of a compound.

Homework

1 Zinc and sodium both form white salts. How can you tell which is which?
2 Find the names and formulae of some compounds used to colour fireworks.
3 Find images of emission spectra for **three** elements which colour flames.

THE HALOGENS

- Vertical columns in the **periodic table** are called **groups**.
- **Elements** in the same group have similar **properties**.
- The **halogens** in group 7 are **reactive** non-metals.
- They react with metals to make salts, for example: iron + chlorine → iron chloride.
- Going down the group, they get less reactive, their colours darken and boiling points increase, for example Cl_2 (green gas), Br_2 (brown liquid), I_2 (grey solid).
- More reactive halogens displace less reactive ones, for example bromine + sodium iodide → sodium bromide + iodine.

 TOP TIP Halogens lose their colour when they form compounds. The colour reappears when they are displaced.

Now try this

b Match each property to the correct halogen.

Chlorine	Displaces bromine
	Grey solid
	Green gas
Bromine	Most reactive
	Has a brown solution
	Brown liquid
	Makes white salts
Iodine	Only displaces iodine

Homework

4 Draw chlorine and sodium bromide solutions before and after they react.
5 Write equations for sodium reacting with each of the halogens.
6 Research possible uses for **two** compounds of each halogen.

OPPOSITE ENDS OF THE TABLE

- The **alkali metals** in group 1 are very reactive.
- They all react with water exothermically:
  sodium + water → sodium hydroxide + hydrogen
- Going down the group, they get more reactive, softer and easier to melt.
- The **noble gases** in group 8 are **inert** but useful.
- Helium's low **density** means it can be used in balloons.
- Neon's red discharge lights advertising signs.
- Argon stops light bulb filaments burning.

 TOP TIP The alkali metals are dangerously reactive but their compounds are vital components of living things, like sodium chloride.

Now try this

c Tick the following statements that are true.

i   Group 1 metals fizz in water ☐

ii   **Na** displaces **H** from water ☐

iii **Li** is more reactive than **Na** ☐

iv   **K** is strong and shiny ☐

v   **KOH** is an alkali ☐

vi   Noble gases are unreactive ☐

vii **He** is an explosive gas ☐

viii **Ne** is a red gas ☐

Homework

7 Prepare a presentation about the reactions of alkali metals.

8 What trends do their melting points and densities show?

9 Find out why sodium and potassium were not discovered until 1907.

PERIODIC PATTERNS

- **Periods** are horizontal rows in the periodic table.
- They let similar elements occupy vertical groups.
- Gaps were originally left for undiscovered elements.
- Group trends allowed their properties to be predicted.
- Elements were originally arranged in order of mass but are now in order of atomic number.
- **Transition metals** are similar and have their own block.
- Copper is a good conductor, which makes it useful for wiring.
- Silver and gold are unreactive, so are good for jewellery.
- Iron is cheap and strong, so is useful for building things like bridges.

Sc	Ti	V	Cr	Mn	Fe	Co	Ni	Cu	Zn
Y	Zr	Nb	Mo	Tc	Ru	Rh	Pd	Ag	Cd
La	Hf	Ta	W	Re	Os	Ir	Pt	Au	Hg
Ac									

Now try this

d Where in the periodic table should each of these elements go?

i   Unreactive metal _____

ii   Reactive non-metal _____

iii Coloured non-metal _____

iv   Unreactive non-metal _____

v   Strong metal _____

vi   Reactive metal _____

vii Dense metal _____

viii Colourless gas _____

 TOP TIP Groups contain families of elements with similar properties.

Homework

10 Use the periodic table to predict the properties of Rb, At and Pt.

11 Design a test on the patterns in the periodic table.

12 Investigate periodic table websites and review the best **two** that you find.

Chemicals everywhere

COMMON CHEMICALS

- Everything is made from **chemicals**.
- Natural chemicals are **purified** before use.
- When pure, 'natural' and 'artificial' versions of **compounds** are identical.
- Common compounds include:
  - ammonia in fertilisers and cleaners
  - carbohydrates like starch in food
  - CO_2 to make drinks fizz and bread rise
  - caustic soda in oven cleaners
  - citric acid to make drinks taste sour
  - ethanoic acid in vinegar
  - hydrochloric acid in the stomach
  - phosphoric acid in cola and toilet cleaner
  - sodium chloride in table salt.

Now try this

a Match the chemicals to their uses.

i	Ammonia	Flavouring lemonade
ii	Strong alkali	Providing energy
iii	Carbon dioxide	Flavouring foods
		Making drinks fizz
iv	Strong acid	Making fertilisers
v	Weak acid	Making bread rise
vi	Carbohydrate	Dissolving grease
viii	Salt	Dissolving limescale

 Acids make foods taste sour.

Homework

1 Produce a poster to show the chemicals in a meal.

2 Prepare a presentation to show **ten** chemicals found in food.

3 Find out how **a)** rock salt is purified and **b)** sodium chloride is made in a laboratory.

MAKING SALTS

- Many useful chemicals are **salts**, including potassium nitrate (oxidiser in fireworks) and ammonium nitrate (fertiliser).
- Soluble salts form when **bases neutralise acids**:
  ACID + solid OXIDE → SALT + WATER
  ACID + solid CARBONATE → SALT + WATER + CO_2
- When the acid is neutralised, the solids stop dissolving.
- To get the salt, **filter** out the excess solid and **evaporate** the water.
- Hydrogen (H) in acids reacts with hydroxide (OH) in alkalis:
  ACID + HYDROXIDE solution → SALT + WATER
- An **indicator** shows when the solution is neutral.

Now try this

b Circle the correct answers.

An acid	HCl	NaCl	NaOH
An alkali	KOH	HNO_3	KCl
A salt	NaOH	H_2SO_4	$NaNO_3$
An oxide	CuO	NaOH	$CuSO_4$
A carbonate	$CaCO_3$	CaO	H_2SO_4
Bases	CuO	$CuCO_3$	NaOH

 Replacing the hydrogen in an acid with a metal gives a neutral salt.

Homework

4 Write step-by-step instructions for making copper sulphate and sodium chloride.

5 Design a test to show which students understand how salts are made.

6 Research the way pH changes during neutralisation.

INSOLUBLE SALTS

- **Pigments** like lead chromate are **insoluble**.
- They can be made by mixing two solutions: the nitrate of their metal and the sodium salt of their non-metal part.

lead nitrate + sodium iodide
↓
lead iodide + sodium nitrate

- In water, the two parts of each soluble salt separate.
- The two parts of the insoluble salt attract each other and form a **precipitate**.
- This is filtered, washed and dried to give the pure salt.

Now try this

c Complete the following equations.

lead nitrate + sodium iodide → _____

barium nitrate + sodium sulphate → _____

silver nitrate + sodium iodide → _____

$AgNO_3 + NaI \rightarrow$ _____

silver nitrate + sodium chloride → _____

$AgNO_3 + NaCl \rightarrow$ _____

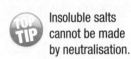 **TOP TIP** Insoluble salts cannot be made by neutralisation.

Homework

7 Lead sulphate is insoluble. Explain how it could be made.

8 Use particle pictures to explain how precipitates form.

9 Research the salt used to paint Van Gogh's sunflowers.

EXTRACTING METALS

- **Metals** are **extracted** from **ores** in the Earth's crust.
- Only the least reactive are found uncombined.
- Many ores are **oxides**, for example iron oxide.
- Adding oxygen is **oxidation**. Taking oxygen away is **reduction**.
- Oxygen is removed to extract metals from oxides.
- Iron, copper and lead can be reduced by carbon:
  carbon + copper oxide → carbon dioxide + copper
- The more reactive the metal, the more stable the ore.
- A blast furnace is needed to extract iron.
- Reactive metals can only be extracted using electricity.

 TOP TIP The easiest metals to extract were the first to be discovered.

Now try this

d Match each metal to an extraction method.

Reduced using electricity	Ag
	Fe
	Pb
Reduced by carbon	Al
	Na
Found as the element	Au

Homework

10 Prepare a presentation to explain oxidation and reduction.

11 Find out how iron is separated from oxygen in a blast furnace.

12 Explain why the same extraction method is not used for every metal.

Everyday chemicals

COOKING CHEMISTRY

- **Thermal decomposition** breaks compounds up.
- Heat decomposes carbonates and hydrogen carbonates. For example:
  calcium carbonate → calcium oxide + carbon dioxide
- Baking powder makes cakes rise because it contains a dry acid and sodium hydrogen carbonate. These react and release bubbles of CO_2:
  sodium hydrogen carbonate + acid
  → sodium salt + carbon dioxide + water
- **Dehydration** removes water and makes food brown.
- Some correlations exist between chemicals and health, but that does not prove they cause the health effects.

 TOP TIP If a solid loses mass when heated, a gas must be escaping.

Now try this

a Match the names and descriptions.

i Carbonate	Contains HCO_3
ii Decomposition	Releases CO_2 if heated
	Breaks up compounds
iii Hydration	Decomposition product
iv Hydrogen carbonate	Addition of water
	Dehydrated starch
v Dehydration	Found in baking powder
vi Carbon dioxide	Loss of water
	Makes food brown
vii Toast	Formula ends in CO_3

Homework

1. Explain how you could prove that roasted limestone decomposes.
2. Research **five** foods that have been linked to health scards or benefits.
3. Prepare a presentation on reactions that occur when food cooks.

GASES

- Testing and labelling **gases**:
  - hydrogen pops with a lit splint
  - oxygen relights a glowing splint
  - carbon dioxide turns **limewater** cloudy
  - ammonia smells of nappies and is **alkaline**
  - chlorine bleaches damp **litmus paper**.
- All gases can be collected in a **gas syringe**.
- Most gases can be collected over water but ammonia is too soluble.
- Upward delivery collects light gases like H_2 and downward delivery collects dense gases like CO_2.
- **Hazard labels** warn users which chemicals are flammable, toxic, corrosive, harmful or an irritant.

 TOP TIP Most gases are colourless and odourless.

Now try this

b Circle the correct answers.

Pops with a lit splint	H_2	CO_2	CO
Lighter than air	CO_2	CO	H_2
Relights glowing splint	CO_2	N_2	O_2
Turns limewater cloudy	H_2	CO_2	N_2
Very flammable	O_2	CO_2	H_2
Bleaches damp litmus	CO_2	NH_3	Cl_2
Heavier than air	N_2	H_2	CO_2
Alkaline gas	NH_3	Cl_2	CO_2
Very soluble	H_2	N_2	NH_3
Smelly gas	N_2	Cl_2	NH_3

Homework

4. Find out which hazard labels these gases should have: NH_3, H_2, Cl_2.
5. Show **four** ways of collecting gases, and the gases each method is best for.
6. Find reactions that could be used to make these gases: O_2, Cl_2, CO_2.

CHEMICALS FROM AIR AND SALT

- Air contains: N_2 (78%), O_2 (21%), Ar (1%), CO_2 (0.04%).
- The mixture condenses when cooled and compressed.
- Solid CO_2 is filtered out.
- N_2 and O_2 are separated out by **fractional distillation**.
- N_2 has the lowest boiling point so it evaporates first.
- Seawater and rock salt are sources of:
  - sodium (Na) for street lamps
  - chlorine (Cl_2) for sterilising water and making PVC
  - sodium chloride for flavouring food and melting ice
  - hydrogen as a fuel, and for making margarine
  - sodium hydroxide to make bleaches, soap and aspirin.

 Distillation separates the gases in air but electrolysis decomposes salt.

Now try this

c Tick the statements that are true.

Air is a mixture of gases	☐
Air decomposes into $O_2 + N_2$	☐
Air's composition is always the same	☐
O_2 and N_2 have different boiling points	☐
O_2 and N_2 are separated by electrolysis	☐
CO_2 solidifies at low temperatures	☐
Sodium chloride is a mixture	☐
Distillation decomposes salty water	☐
H_2, Cl_2 and NaOH are made from salt	☐

Homework

7 Find out what O_2 and N_2 are used for.

8 Prepare a presentation about products made from salt.

9 Explain how air is liquefied and how fractional distillation works.

ALTERNATIVE FUELS

- **Fossil fuels** contribute to the **greenhouse effect**.
- **Biofuels** are **renewable** and 'greenhouse neutral', but growing them uses up farmland, and cars need to be modified to use them.
- **Ethanol** is made from sugar cane or sugar beet.
- **Hydrogen fuel** is pollution free – it only makes water.
- An energy source is needed to extract H_2 from water.
- Storing hydrogen is difficult because it is flammable.

Now try this

d Match each fuel to the right descriptions.

Hydrogen	Releases CO_2 and H_2O
	Greenhouse-neutral
Biodiesel	Difficult to store
	Used in fuel cells
	Made from plant oils
Ethanol	Pollution-free
	Releases H_2O only
Fossil fuel	Made from sugar

 Hydrocarbon fuels release CO_2 and H_2O when they burn.

Homework

10 Research biodiesel, hydrogen or ethanol and present its positive and negative points.

11 Search the Internet for cars designed to run on each of the fuels mentioned in Homework question **10**.

12 Some vehicles have hydrogen fuel cells. Find out how these work.

Fuels and air quality

BURNING OIL

- **Fractional distillation** separates **crude oil** into:
  - gases for portable cookers (top of **tower**)
  - petrol for cars
  - kerosene for aircraft
  - diesel oil for cars and lorries
  - naphtha for synthesising organic chemicals
  - fuel oil for ships
  - bitumen for road surfaces (bottom of tower).
- Going down the tower:
  - molecules get bigger and harder to ignite
  - **boiling points** and **viscosity** increase.

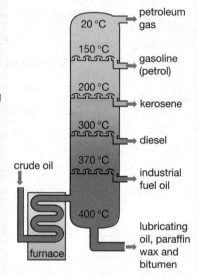

crude oil

20 °C → petroleum gas

150 °C → gasoline (petrol)

200 °C → kerosene

300 °C → diesel

370 °C → industrial fuel oil

400 °C → lubricating oil, paraffin wax and bitumen

furnace

Now try this

a Are the following **small** or **large** hydrocarbon molecules?

Low boiling point _____

Goes to top of tower _____

C_4H_{10} _____

Used to surface roads _____

Hard to vaporise _____

Very viscous _____

Used in petrol _____

Easy to light _____

$C_{30}H_{62}$ _____

 TOP TIP Fractions are not pure but contain hydrocarbons with similar numbers of atoms.

Homework

1 Make a presentation to show a use for each oil fraction.

2 Show how short and long molecules separate during distillation.

3 Find **three** chemicals that are added to petrol and outline their roles.

BURNING METHANE

- Useful fuels:
  - burn completely producing no soot or ash
  - give out maximum heat energy per gram.
- In faulty gas heaters, **incomplete combustion** gives:
  - sooty deposits of carbon (C)
  - toxic **carbon monoxide** (CO).
- Incomplete combustion releases less energy.
- CO reduces the blood's ability to carry oxygen.

 TOP TIP **Hydrocarbon fuels** usually make water vapour and CO_2. Oxygen shortages produce CO or C instead.

Now try this

b Finish these equations.

methane + oxygen → _____

$CH_4 + 2O_2 \rightarrow 2H_2O +$ _____

methane + less oxygen → _____

$CH_4 + 1\frac{1}{2}O_2 \rightarrow 2H_2O +$ _____

methane + much less oxygen → _____

$CH_4 + O_2 \rightarrow 2H_2O +$ _____

Homework

4 Design a safety leaflet to warn gas boiler owners about carbon monoxide.

5 Animate a PowerPoint presentation to show methane reacting with oxygen molecules.

6 Find out how a Bunsen's oxygen supply affects the burning methane.

See pages 166–179 of Collins GCSE Science

AIR QUALITY

- **Pollution** can be determined by direct measurements or from the distribution of sensitive plants and animals.
- Websites have data about **global warming** and **acid rain**, but need to be checked for authenticity and bias.
- Changing pollution levels can be plotted on graphs.
- If the number of asthma cases shows the same trend, pollution could be the cause, but it might not be.
- Correlations do not prove one thing causes another.
- Pollution makes asthma worse but there is no clear evidence that it causes it.

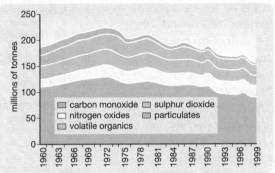

carbon monoxide □ sulphur dioxide
nitrogen oxides □ particulates
volatile organics

Now try this

c Fill in the missing words.

The graph shows that air pollution was _____ until 1972. Since then, the overall trend has been a _____ in all pollutants except nitrogen oxides. Asthma cases have risen so they _____ be caused by air pollution.

 To prove one thing causes another you need to find out how it does it.

Homework

7 Find out how catalytic converters improve air quality.

8 Find out how air quality is monitored in your nearest city.

9 What can lichens tell us about air pollution?

CHANGING THE ATMOSPHERE

- Our **atmosphere** has changed since Earth formed:
  - volcanoes released CO_2, ammonia and steam
  - oceans condensed and CO_2 dissolved in them
  - **photosynthesis** released O_2
  - O_2 released N_2 from ammonia.
- Average temperatures and CO_2 levels are rising due to **deforestation** and burning **fossil fuels**.
- The theory that **global warming** is due to increased CO_2, has been widely accepted.
- Predictions about global warming are based on computer models, which may not be reliable.
- Following the precautionary principle means cutting CO_2 emissions now, rather than waiting for firmer evidence about global warming.
- **Recycling** saves resources and cuts the energy and CO_2 emissions needed to make new products.

 Short-term fluctuations may mask long-term trends.

Now try this

d Decide if these **increase (↑)** or **decrease (↓)** atmospheric CO_2.

Deforestation _____
Burning petrol _____
Recycling _____
Drought _____
Forest fires _____
Renewable fuels _____
Insulating homes _____
Growing populations _____
Coal-fired power stations _____

Homework

10 Prepare a presentation on how the Earth's atmosphere has changed.

11 Find out what 'environmental footprints' are, and why they are useful.

12 Design a leaflet to explain 'sustainable development' to Year 7 students.

New materials

BETTER BY DESIGN

- **Smart materials** with useful properties include:
  - Lycra® for sports clothes (keeps its elasticity)
  - Thinsulate™ for ski clothes (traps air for insulation)
  - carbon fibres for sports equipment (tough and light)
  - PTFE for artificial hip joints (a low-friction polymer).
- Materials that change in response to a stimulus include **memory alloys** (regain their shape when heated) and **polymers** (shrink when they absorb water).
- **Nanotechnologies** work with clusters of atoms.
- **Nanoparticles** give materials unique properties.
- Titanium dioxide nanoparticles make sunscreens absorb harmful ultraviolet light but look transparent.
- Sensational media stories about nanotechnology make people worry that problems will arise.

Now try this

a Match each material to a product.

i	Sledge runners	carbon fibre
ii	Self-repairing cars	Lycra®
iii	Sleeping bags	PTFE
iv	Sportswear	Thinsulate™
v	Racing car body	shape memory alloy

 TOP TIP Materials with changeable properties can be used as sensors.

Homework

1 Find out how smart food wrappings change colour.
2 List the properties of each material in a snow boarder's kit.
3 How are PVA gels able to move when they get wet?

POLYMERS

- Teflon® and the glue on 'Post-it' notes were created by accident – uses were found later.
- In GORE-TEX®, tough **polymers** are layered over a fragile layer of waterproof PTFE, to protect it. It lets sweat escape but keeps out rain drops.
- Kevlar® is tough, lightweight and flame-resistant.
- New products are designed to have **properties** that will suit their intended usage.

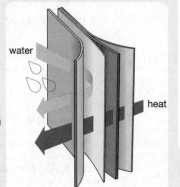

water

heat

Now try this

b Match each property of Kevlar® to a use.

i	Withstands tension	Glass workers' gloves
ii	Resists impacts	Fire fighter's uniform
iii	Lightweight	Parachute ropes
iv	Hard-wearing	Bullet-proof vests
v	Flame-resistant	Safety helmets
vi	Cut-resistant	Run-flat tyres

 TOP TIP Different materials can be laminated together to give the best overall properties.

Homework

4 Find out how Teflon™ was discovered.
5 Prepare a presentation on the uses of Teflon (PTFE).
6 Research **three** examples of smart materials.

ALCOHOL

- Beer and wine are made by **fermentation** of sugar using yeast enzymes:

sugar → ethanol + carbon dioxide

- **Ethanol** lengthens reaction times and increases the likelihood of accidents.
- The more alcohol in the blood, the longer it takes to spot hazards.
- Heavy drinking causes cirrhosis of the liver.

 TOP TIP Enzymes in the liver break down ethanol but this takes time.

Now try this

c Tick the facts about alcohol that are true.

i	It circulates in the blood	☐
ii	It prevents dehydration	☐
iii	It is made from sugar	☐
iv	It is made by yeast	☐
v	It stimulates the brain	☐
vi	It slows your reactions	☐
vii	It is good for your liver	☐

Homework

7 Produce a poster on the possible consequences of excess alcohol consumption.

8 Prepare a presentation on the alcohol concentrations in popular drinks.

9 Find out how the police test for alcohol in blood and urine.

MIXING AND SEPARATING

- Many useful materials are **mixtures**.
- **Emulsions** contain **immiscible liquids**.
- Mayonnaise has oil drops dispersed in water.
- **Emulsifiers** stop emulsions separating into layers. Their **hydrophobic** ends stick into the oil and **hydrophilic** ends face into water.
- **Oxidation** spoils food.
- Water lets **microbes** grow.
- Intelligent packaging removes water or prevents reactions with oxygen.

Now try this

d Match the words and definitions.

i	Oxidation	Water loving
ii	Emulsify	Cannot be mixed
iii	Emulsifier	Disperse one liquid in another
iv	Hydrophilic	Stops emulsions separating
v	Water	A reaction with oxygen
vi	Hydrophobic	Water hating
vii	Immiscible	Lets microbes grow

 TOP TIP **Antioxidants** preserve food by slowing its reactions with oxygen.

Homework

10 Highlight the oil, water and emulsifier in a mayonnaise recipe.

11 Prepare a presentation to explain why food spoils.

12 Research the advantages of smart packaging.

Measuring electricity

CURRENT, VOLTAGE AND RESISTANCE

- Electrical **current** is the flow of charged particles (**electrons** or **ions**).
- Current in a **series circuit** is always the same.
- Current in a **component** is measured using an **ammeter** connected in series.
- Current is measured in **amperes** (or amps or A).
- The **voltage** or **potential difference** across a component is measured using a **voltmeter** placed in **parallel**.
- Voltage is measured in **volts** or V.
- **Resistance** is measured in **ohms** or Ω.
- The voltage V across a component is related to the current I in the component and its resistance R by the equation: $V = I \times R$.

Now try this

a Complete the sentences by filling in the gaps.

A 50 Ω resistor and a lamp are connected in series. The current in the circuit is 0.20 A. The _____ in each component will be the same and equal to _____. The voltage across the resistor may be measured using a _____ placed in parallel. The voltage across the resistor is _____ V.

TOP TIP The size of the current in a circuit depends on its resistance. The current is smaller for a larger resistance.

Homework

1 Make a list of all the electrical quantities and their units.
2 With the help of circuit diagrams, describe how you would use an ammeter and a voltmeter in a circuit.

RESISTORS AND FILAMENT LAMPS

- The **resistance** of a fixed **resistor** remains constant for all values of current.
- The current-voltage graph for a resistor is shown right.
- The resistance of a **filament lamp** increases with increasing current because of heating.
- The current-voltage graph for a filament lamp is shown right.

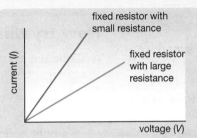

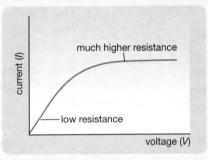

Now try this

b Find the following key words in this word search.

current   voltage   resistor
lamp   graph

A	V	C	L	A	M	P	D
Q	O	U	V	R	N	G	G
Z	L	R	G	L	B	J	P
X	T	N	E	R	R	U	C
F	A	O	K	T	A	P	O
G	G	F	U	E	E	P	P
H	E	J	K	U	Y	T	H
R	E	S	I	S	T	O	R

Homework

3 Describe the electrical properties of a resistor.

4 Make a list of devices that have resistors.

5 Describe the electrical properties of a filament lamp.

SENSING LIGHT AND HEAT

- The **resistance** of a **light-dependent resistor** (LDR) decreases as the brightness of light falling on it increases.
- LDRs are used in burglar alarms and for automatically operating street lights.
- LDRs are also used in digital cameras to automatically control the amount of light entering the camera.
- The resistance of a **thermistor** ——▭—— decreases as its temperature increases.
- Thermistors are used as safety devices in fridges, projector lamps and computers.

TOP TIP You can find out what happens to the current in a circuit by using: $I = \dfrac{V}{R}$.

Now try this

c Tick the correct statements.

Thermistors are used in controlling street lights. ☐

An LDR is connected to a cell. The circuit current decreases when it gets darker. ☐

A thermistor is connected to a cell. The circuit current increases when it gets hot. ☐

Large current in a thermistor can decrease its resistance due to heating. ☐

Homework

6 Draw the electrical symbols of all the components from this section.

7 Design a circuit to monitor the brightness of light in a room.

8 Design a circuit to monitor the temperature in a greenhouse.

Producing electricity

CELLS AND BATTERIES

- A **battery** consists of several chemical **cells** joined together in **series**.
- A cell or a battery produces a **direct current** (d.c).
- A **solar cell** changes light energy into electrical energy. It too produces direct current.
- The direction of direct current remains the same.
- **Electrons** in a circuit travel from the negative terminal to the positive terminal.
- Conventional current travels from the positive terminal to the negative terminal.
- There are basically two types of chemical cells: dry (or non-rechargeable) cells and rechargeable cells.
- The metals used in chemical cells are toxic and can have harmful effects:
  - cadmium can cause kidney problems, affects our bones and can cause cancer
  - mercury can cause miscarriages
  - lead can damage our nervous system.
- The capacity of a cell or a battery is quoted in Amp-hours. This can be used to predict how long a battery will last. For example, a 20 Amp-hours battery can produce a 1 A current for 20 hours or a 0.5 A current for 40 hours, etc.

Now try this

a Circle the correct answer for each statement.

i A solar cell changes light energy into:

    heat   light   electrical energy

ii The size of a direct current can change but its direction:

    does change   does not change

iii A 60 Amp-hours battery supplying a current of 2 A will last:

    120 hrs   60 hrs   30 hrs

iv The metals used in chemical cells must be disposed of carefully as they are:

    hot   toxic   explosive

Homework

1 Make a list of **five** devices that use cells or batteries. For **one** of the devices, explain why having a battery is helpful.

2 Describe the energy changes taking place in a solar cell and a chemical battery.

3 Write a short paragraph on the dangers of chemical cells.

DYNAMOS

- Another term for '**potential difference**' is '**voltage**'.
- **Electromagnetic induction** refers to the creation of voltage across the ends of a wire or coil when it is moved in a **magnetic field**.
- The size of the **induced** voltage across the ends of a wire is bigger when it moves faster through a stronger magnetic field.
- Changing the direction of the motion of the wire changes the direction of the induced voltage or current.
- In a bicycle **dynamo**, a magnet is rotated near a coil. The induced voltage repeatedly changes direction. The current produced by a dynamo is **alternating current** (a.c.).
- Rotating the dynamo magnet faster increases the size of the induced voltage and its frequency. (The induced voltage is directly proportional to the frequency.)

 TOP TIP In order to induce a current or a voltage you have to transform **kinetic energy** into **electrical energy**.

Now *try this*

b Match the beginnings and endings to make complete sentences.

Beginning	Ending
i A dynamo produces an	coil.
ii To induce a current in a wire you have to move it in a	frequency of the current.
iii In a dynamo the magnet rotates near a stationary	magnetic field.
iv Rotating the coil of a dynamo faster increases the	alternating current.

Homework

4 In your own words, describe electromagnetic induction.

5 Describe the factors that affect the output current from a dynamo.

IMPACT OF ELECTRICITY

- **Superconductors** are materials that have very low **resistance**.
- The Maglev trains move along by **electromagnets**.
- Making the electromagnets from superconductors reduces heat losses and therefore saves energy.
- The telephone system has had a huge impact on modern society.
- Modern computers have electrical components and devices within them.
- The Internet uses computers to process information at high speed.

Now *try this*

c Circle the correct answers.

  i The resistance of superconductors is quite

  small   large

  ii Using superconductors can save

  energy   current

  iii Computers process information very

  slow   fast

  iv The Maglev trains work using

  dynamos   electromagnets

Homework

6 The Internet has changed the world in many ways. Write a short paragraph to describe how the Internet has helped you.

7 How do superconductors differ from ordinary conductors? Why do we not have them in our homes?

Being in charge

GENERATORS AND MOTORS

- A simple **generator** consists of a coil rotating in a magnetic field.
- A **voltage** is **induced** across the ends of the coil as it 'cuts' the magnetic field.

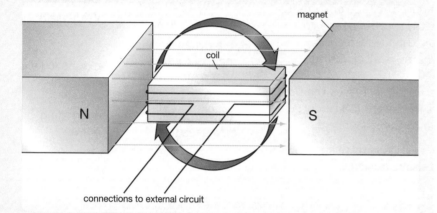

- Rotating the coil faster in a stronger magnetic field can increase the output voltage from a generator.
- A **turbine** is used to rotate the coil or the magnet.
- For fossil and nuclear fuels, the turbine is made to rotate by high-pressure steam.
- The turbine can be rotated directly by sources such as wind, wave, tidal and hydroelectric.
- The **non-renewable** energy resources are: fossil fuels (coal, gas, oil and peat) and nuclear.
- The **renewable** resources are: biofuels, hydroelectric, tidal, wave and wind.
- The **National Grid** transports electricity from power stations to our homes and industry.
- A current-carrying wire is surrounded by a magnetic field.
- A current-carrying wire placed in a magnetic field experiences a force. The direction of this force may be found using **Fleming's left-hand rule**:
  - thu**M**b → **M**otion
  - **F**irst finger → **F**ield
  - se**C**ond finger → **C**urrent.
- An electric motor has a rectangular coil placed in a magnetic field.

Now try this

a Fill in the missing words.

In a coal-burning power station, high-pressure _____ is used to turn the turbine. The turbine then rotates either the _____ or the _____. This induces a voltage across the ends of the _____. In the UK, the electricity is transported at high voltages using the National _____.

 TOP TIP A generator changes kinetic energy into electrical energy and a motor does the opposite.

Homework

1 Draw a block diagram to show how electricity is generated at a power station.
2 Generators and motors are often confused. Write a short paragraph to describe how they differ.

EFFICIENCY AND POWER

- The **efficiency** of a device is defined as:

$$\text{efficiency} = \frac{\text{useful output energy}}{\text{total input energy}} \times 100\ \%$$

- A lamp is supplied with 30 J of electrical energy. 28 J is lost as heat and 2 J is transformed into light. The efficiency of the lamp is:

$$\text{efficiency} = \frac{2}{30} \times 100\ \% = 6.7\ \%$$

- Electrical **power** is defined as the rate of transfer of electrical energy:

$$\text{power} = \frac{\text{energy transferred}}{\text{time taken}}$$

- Power is measured in **watts** (W) or **joules per second** (J/s).
- The electrical power is related to voltage and current by the equation:

$$\text{power} = \text{voltage} \times \text{current} \qquad P = VI$$

- 1 kWh is the energy transformed by a 1 kW device working for a time of 1 hour.
- 1 kWh = 3 600 000 J.
- The cost of electricity is calculated using:

$$\text{cost} = \text{power (in kW)} \times \text{time (in hours)} \times \text{cost of 1 kWh}$$

- Reducing energy losses from our homes can reduce energy costs.

Now try this

b Tick the correct statements.

Power is measured in joules. ☐

The kWh is an alternative unit for power. ☐

The energy transformed by a 2 kW heater working for 3 hours is 6 kWh. ☐

A motor can have an efficiency of 120 %. ☐

Homework

3 Define the kilowatt-hour and explain why it is a convenient unit of energy.

4 List the power rating of **four** mains-operated appliances in your home. For each appliance determine the energy transformed in 1 hour (3600 seconds).

SAFETY

- A mains plug has three important connections: **earth**, **neutral** and **live**.
- The **fuse** is a safety device attached to the live terminal. A large current will melt the fuse and cut off the supply.
- Metal appliances are earthed to protect you from accidental shocks.
- If an earthed appliance accidentally becomes live, the current travels to ground through the earth cable rather than the person who touches the appliance.
- **Residual current circuit breakers** (RCCB) are also safety devices.

 TOP TIP RCCB cut off the supply more quickly than fuses and can also be reset and reused.

Now try this

c The diagram shows a mains plug. Label the earth, live and neutral terminals and the fuse.

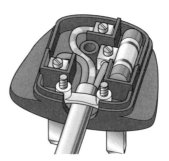

Homework

5 Describe how a fuse works.

6 Explain why plastic-cased appliances do not need to be earthed.

It's all about waves

UNDERSTANDING WAVES

- All **waves** carry **energy** from one place to another.
- There are two types of waves: **longitudinal** and **transverse**.
- A longitudinal wave has vibrations parallel to the direction of the wave velocity.
- A transverse wave has vibrations 90° to the direction of the wave velocity.
- Examples of transverse waves are: ripples on the surface of water, all electromagnetic waves (for example, light), and seismic S-waves.
- Examples of longitudinal waves are: sound, ultrasound and seismic P-waves.
- **Seismic waves** provide detailed information about the interior structure of our Earth.
- The **amplitude** of a wave is the maximum distance of the wave from its undisturbed position. Amplitude is measured in metres (m).
- The **frequency** of a wave is the number of waves produced per unit time. Frequency is measured in **hertz (Hz)**.
- The **wavelength** of a wave is the distance between two neighbouring peaks. Wavelength is measured in metres (m).
- The speed v of the wave is related to its frequency f and the wavelength λ by the equation: $v = f\lambda$. The speed is measured in metres per second (m/s).
- The speed of a wave is given by:

$$\text{speed} = \frac{\text{distance travelled}}{\text{time}}$$

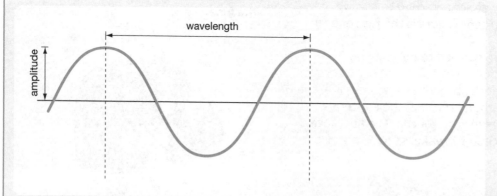

Now try this

a Circle the correct answers.

   i The speed of a wave is measured in:

   m   m/s   Hz

   ii Wavelength and amplitude can be measured in:

   m   m/s   Hz

   iii Frequency is measured in:

   m   m/s   Hz

   iv Another name for seismic waves is:

   light   earthquake   wind

Homework

1. In your own words define amplitude, wavelength, frequency and speed of a wave.
2. Describe the similarities and differences between longitudinal and transverse waves.
3. Use the Internet to find some images of the destruction caused by earthquakes. Make a list of the websites for the class.

REFLECTION, REFRACTION AND TIR

- For a wave **reflected** at a smooth surface, the **angle of incidence** is equal to the **angle of reflection**.
- In **echo-sounding**, the depth of the water is given by:

  $$\text{depth} = \frac{\text{speed of wave} \times \text{delay time}}{2}$$

- **Refraction** is the bending of a wave as it travels between mediums.
- Refraction occurs because the speed of the wave changes as it travels.
- When a ray of light travels from glass (or water) into air:
  - small angles: the ray is refracted and reflected at the boundary
  - incidence angle equal to the **critical angle**: the ray is refracted along the boundary and there is a weak reflection
  - angles greater than the critical angle: there is no refraction but a strong **total internal reflection (TIR)** takes place.
- TIR is used in prisms, bicycle reflectors and **optical fibres** (endoscope).
- Modern telecommunications systems use optical fibres to transmit data at high speed over long distances.

Now try this

b Is each statement **true** or **false**?

A shiny mirror will reflect light. _____

During refraction, the speed of the wave changes. _____

At the critical angle, a ray of light shows no refraction. _____

The light in an optical fibre shows total internal reflection. _____

The bending of light is called reflection. _____

TOP TIP TIR cannot occur if light travels into an optically denser medium (for example, from air into water).

Homework

4 With the help of diagrams, describe critical angle and total internal reflection.

5 Use the Internet to research and write about the applications of optical fibres.

6 Draw the path of rays in a prism and a bicycle reflector.

ELECTROMAGNETIC WAVES

- The **electromagnetic waves** in order of decreasing wavelength (or increasing frequency) are: **radio waves**, **microwaves**, **infrared** (IR), **visible light**, **ultraviolet** (UV), **X-rays** and **gamma rays**.
- All waves in the **electromagnetic spectrum** are **transverse**, have different wavelengths or frequencies but travel at the same speed (300 000 km/s) in a vacuum.
- Radio waves do not harm the human body.
- Microwaves can heat human tissues.
- Infrared radiation can burn skin.
- Intense visible light (for example, lasers) can harm the eyes.
- Ultraviolet radiation can cause sunburn and cancer.
- X-rays and gamma rays can mutate cells and cause cancer.

Now try this

c Fill in the missing words.

Radio waves are the same as X-rays because they are both _____ waves. Both waves can travel through a _____ at the same _____. X-rays have _____ wavelength than radio waves and they are much more harmful to humans because they can cause _____.

Homework

7 Devise a mnemonic to help you remember the order of the waves in the electromagnetic spectrum.

8 Write a short paragraph on the dangers of electromagnetic waves to humans.

Uses and dangers of waves

DANGERS OF WAVES

- **Radio waves** cause no damage to humans.
- **Microwaves** can cause internal heating of body tissues.
- Mobile phone systems use low power microwaves. Most scientists think they pose very small health risks to humans.
- **Infrared** causes heating of body tissues.
- Over-exposure to **ultraviolet radiation** can lead to sunburn and skin cancer.
- There are three types of ultraviolet radiation: **UVA**, **UVB** and **UVC**.
- UVA has the longest wavelength or the lowest frequency. It causes early ageing of the skin, DNA damage and cancer.
- UVB has a shorter wavelength than UVA. It does the same damage as UVA but it also stimulates the production of vitamin D in our bodies.
- UVC has the largest frequency. Most of this is stopped by the ozone layer. It is the more dangerous of the ultraviolet radiations.
- **X-rays** and **gamma rays** mutate and destroy body cells and can cause cancer.

 TOP TIP Larger amplitude and higher frequency radiations cause more damage to human tissues.

Now try this

a Tick the correct column for each radiation.

	No damage	Heating	Sunburn	Cancer
Radio waves	☐	☐	☐	☐
Microwaves	☐	☐	☐	☐
Infrared radiation	☐	☐	☐	☐
Ultraviolet radiation	☐	☐	☐	☐
X-rays	☐	☐	☐	☐
Gamma rays	☐	☐	☐	☐

Homework

1. Describe the electromagnetic waves that have no harmful effect on humans.
2. Describe the properties of the three types of ultraviolet radiations.
3. Write a short paragraph to describe the effect of X-rays and gamma rays on healthy cells.

USES OF WAVES

- There are two scanning techniques: scanning by reflection and scanning by absorption.
- Iris recognition works on the basis of reflection of light. The pattern of the iris is unique.
- Reflection of **ultrasound** is used to form detailed foetal images.
- **X-rays** are absorbed by dense bone but not soft tissue. X-ray photographs can be used to identify broken bones.
- Water molecules absorb **microwaves** of particular wavelengths. Satellite images are used to monitor the amount of microwaves absorbed by clouds.
- **Ultraviolet** radiation is used to detect forged banknotes by fluorescence.
- Seismic **S-waves** and **P-waves** are used to study the internal structure of the Earth.

Now *try this*

b Find the following key words in this word search.

reflect   scan   iris
absorb   foetus
bone   notes

I	P	A	H	R	A	D	Y
X	R	T	G	E	N	O	B
A	K	I	N	F	T	P	F
B	A	V	S	L	E	N	N
S	V	V	I	E	E	A	O
O	A	L	S	C	N	C	T
R	F	O	E	T	U	S	E
B	N	E	E	T	U	Y	S

Homework

4 Draw a mind map for scanning by reflection.

5 Draw a mind map for scanning by absorption.

DIGITAL OR ANALOGUE?

- An **analogue signal** can have an infinite number of values between any two levels. A microphone picking up sound shows an analogue trace on an **oscilloscope**.
- A **digital signal** can only have two possible values. It is made up of 1s and 0s.
- Computers use digital signals to perform operations at high speed.
- Analogue signals are affected by external electrical noise.
- Digital signals can be sent over long distances along cables and are not affected by external electrical noise.
- Digital information can be stored indefinitely on CDs and DVDs.
- The Internet can be used to download music in digital form.
- Digital signals are used in instruments like the synthesiser.

Now *try this*

c Are the following statements **true** or **false**?

Sound is an analogue signal. _____

An example of a digital signal is 123545. _____

A CD stores digital information. _____

Digital signals can be transmitted at high speed. _____

Homework

6 Make a list of devices in the home or office that use digital signals.

7 Describe how digital signals have affected the music industry.

Space

THE SPACE AROUND US!

- Our **Solar System** consists of the **Sun**, nine **planets** and their accompanying **moons**, **asteroids** and **comets**.
- Pluto – furthest from the Sun – is the coldest planet.
- The closest planets – Mercury and Venus – are extremely hot.
- The planets have different masses and hence different surface **gravitational fields**.
- The planets have their own atmospheres, mostly unsuitable to humans.
- The planets and asteroids **orbit** the Sun in elliptical orbits. (In fact, the orbits are almost circular).
- The comets have much more distinct elliptical orbits. They spend most of their time far from the Sun.
- There is a small chance that the Earth may be hit by a comet in the future. This would cause great devastation.
- Our Sun is one of the billion **stars** in our spiral **galaxy** called the **Milky Way**.
- The **Universe** contains millions of galaxies. The space between galaxies is empty space (vacuum).
- Other than our Sun, our nearest star is about 4.2 light years away. Our nearest galaxy is Andromeda, which is 2.2 million light years away.
- The following objects are in order of increasing size: comet, asteroid, moon, planets, Sun, galaxy, Universe.

Now try this

a Are the following statements **true** or **false**?

Comets have elliptical orbits. _____

All the planets have the same temperature. _____

Our Sun is just a star. _____

Our nearest galaxy is closer than our nearest star. _____

There are stars between galaxies. _____

 TOP TIP Our Sun is just an average star in the Universe.

Homework

1. Describe the content of the Universe.
2. Using a search engine, search for images of distant galaxies. Make a list of **three** websites for the class.
3. "Our Solar System is very small compared with the size of the Universe." Write a short paragraph to support this statement.

THE PHYSICS OF SPACE

- The **mass** of an object is a measure of the amount of matter it contains. For a given object, its mass in kilograms is always the same.
- The **weight** of an object is the **gravitational force** acting on it. Weight is measured in **newtons** (N).
- The force F acting on an object of mass m with an acceleration a is given by:
$$F = ma$$
- weight (N) = mass (kg) × acceleration of free fall (m/s²)
weight (N) = mass (kg) × gravitational field strength (N/kg)
$$W = mg$$
- When two objects interact, each exerts an equal but opposite force on the other. 'Action' = 'Reaction' (Newton's third law).

Now *try this*

b Take the Earth's gravitational field strength g to be 10 N/kg. Circle the correct answers.

i The mass of an object is always:

the same   different

ii A 2.0 kg object on Earth has a weight of:

2.0 N   20 N

iii The force on a 4.0 kg object with acceleration 2.0 m/s² is:

8.0 N   0.5 N

Homework

4 Describe to a GCSE student the difference between mass and weight.

5 Describe how a rocket propels itself in space.

EXPLORING SPACE

- In interplanetary space, there is no air, the temperatures are extremely low and there is a lack of **gravity**, which gives rise to weightlessness.
- For long journeys in space, astronauts have to exercise to prevent muscle wastage and depletion of calcium in their bones.
- A spaceship carrying astronauts has to produce electricity using solar panels or fuel cells, create artificial gravity (by rotating the spaceship), have an adequate supply of air, have heating and cooling systems, prevent damage to the spaceship from micrometeorites and protect the astronauts from the dangers of radiation from space and the Sun.
- Scientists have gathered information from other planets and our Moon by using **landers** to collect soil samples or **fly-by probes** to collect information about composition of the atmosphere and surface.
- The Search for Extraterrestrial Intelligence (SETI) is happening with radio telescopes scanning the sky for messages from intelligent extraterrestrial life.

Now *try this*

c Match the beginnings and endings to make complete sentences.

Beginning	Ending
Interplanetary space	prevent muscle wastage.
Astronauts have to exercise regularly to	has no air.
Landers and fly-by probes	on other planets in the Universe.
There is a small chance of extraterrestrial life	are unmanned spacecrafts.

Homework

6 How has the exploration of space been beneficial to us?

7 Outline the advantages and disadvantages of exploring planets using landers and fly-by probes rather than spaceships with astronauts.

8 Sketch an image of an extraterrestrial being and explain the choice of features.

Stars and the Universe

A STAR IS BORN

- **Stars** are formed from clouds of gas (hydrogen) and dust in space.
- **Gravitational forces** pull the gases and dust particles together – known as **gravitational collapse**. This increases the temperature.
- **Fusion** reactions between **hydrogen nuclei** start when the temperatures become too high. Energy is released in the form of heat and light. A star is born.
- Within a star's core, hydrogen nuclei fuse together to form **helium nuclei**.
- A star has a stable size because of the balance between the attractive gravitational force and expansive forces created by the fusion reactions.
- After billions of years, when the hydrogen 'fuel' runs out, the star begins to die.

Now try this

a Circle the correct answer for each statement.

i The temperature of a gas cloud increases as the cloud shrinks due to:

   gravity   fusion

ii A star releases energy because of nuclear reactions called:

   kinetic   fusion   gravity

iii A star can last for:

   seconds   days   billions of years

Homework

1 In your own words, describe what is meant by 'a star'.
2 Draw a block diagram to illustrate how a star is formed.
3 Describe what is meant by fusion. Why are high pressures and temperatures needed to start these reactions?

THE DEATH OF A STAR

- When the hydrogen fuel runs out, fusion reactions between helium nuclei start to occur.
- For a star of mass similar to our Sun, the star expands to a **red giant**, which has a cooler surface and appears red in colour.
- The outer layers of the red giant are gently blown away in the form of a **planetary nebula** and its core contracts to form a **white dwarf**, which eventually turns into a **brown dwarf** and then a **black dwarf**.
- A star more massive than our Sun expands to a **red supergiant** when the hydrogen fuel runs out.
- When the nuclear reactions stop, the core of the star suddenly collapses and then rebounds against the dense inner core, ejecting dust and gas into space. This explosive star is known as a **supernova**.
- The star left behind is a very dense **neutron star**.
- If the original star is extremely massive, when it collapses it leaves behind a **black hole**.
- A black hole is dense and creates strong gravitational forces around it – not even light can escape!

Now try this

b A star is close to its death. Arrange the following stages in the correct sequence.

For a star like our Sun: red giant, star, planetary nebula, black dwarf, white dwarf

_____

_____

_____

For a star more massive than our Sun: black hole, star, red supergiant, supernova

_____

_____

_____

Homework

4 Describe the fate of a star like our Sun.

5 Draw a block diagram to illustrate the fate of a star much more massive than our Sun.

6 In your own words, describe a red giant, a supernova, a black hole and a white dwarf.

THE MODELS OF THE UNIVERSE

- The Universe began from a sudden expansion of space and time some 15 billion years ago. This event is known as the **Big Bang**.
- The evidence that the Universe is expanding comes from: the observation that all galaxies are moving away from each other; the existence of cosmic microwaves; the presence of large amounts of helium and hydrogen in space.
- The expansion of the Universe has cooled it to its current temperature of about –270 °C.
- The spectrum of light from all galaxies moving away from us shows **red shift** (the entire spectrum is moved to longer wavelengths).
- The more distant galaxies are moving away from us much faster than the closer ones.
- If there is not enough matter in the Universe, it will expand forever.
- If there is enough matter in the Universe, gravity will slow down the expansion rate and eventually start to contract the Universe towards a **Big Crunch**.
- In an oscillatory Universe, the Universe repeatedly expands and contracts.
- It is difficult to know how much matter there is in the Universe because of the existence of **dark matter** (dead stars, black holes, neutrinos, etc).
- In the rejected model of the steady state theory of the Universe, the Universe had no beginning and will have no end.

Now try this

c Match the beginnings and endings to make complete sentences.

Beginning	Ending
The Universe began from an event known as	the Universe is expanding.
All galaxies are moving away from each other because	red shift.
The light from all galaxies shows	it will expand forever and get cooler.
If there is not enough matter in the Universe	the Big Bang.

Cosmic microwave is the leftover radiation from the Big Bang.

Homework

7 List all of the evidence that supports the Big Bang model of the Universe.

8 Write a short paragraph to describe the fate of the Universe.

9 Use a search engine like, to search for images of 'red shift'. Make a list of **three** websites for the class.

Inside living cells

GENETIC INFORMATION

- **DNA** is made up of four different **bases** – **thymine, guanine, adenine** and **cytosine**.
- Sections of DNA called **genes** code for specific **proteins**. These result in certain traits.
- Proteins are made up of **amino acids**. There are many different types of amino acids. The order and type of amino acid determines the structure and type of protein produced.
- Each amino acid is coded for by three bases. If one of these bases is mutated, this can result in the incorrect amino acid being coded for and consequently the wrong protein being made. This is known as a **mutation**.

 You can think of chromosomes as a recipe book – they tell the body how to make all the characteristics. Genes are like individual recipes, they code for specific characteristics.

Now try this

a Fill in the missing words.

There are _____ different bases that make up DNA. _____ bases code for _____ amino acid. Many _____ _____ make up proteins. If DNA undergoes a _____, this results in the incorrect _____ _____ being coded for. If the _____ _____ are not in the correct sequence, the wrong _____ might be made. This results in the characteristic being affected.

Homework

1 Draw a flow diagram explaining how the order of specific bases on DNA results in certain proteins being produced.

2 How can the mutation of one base result in the incorrect protein being made?

3 Which genetic mutation causes cystic fibrosis? Which protein is affected?

GENES

- Sections of **DNA** which code for certain **characteristics** are known as **genes**.
- Through the human genome project, we have identified all the genes found on all 23 pairs of **chromosomes**.
- Using special **enzymes**, we are able to 'cut out' useful genes and place them into a virus.
- We can then use this virus to infect bacteria cells.
- The new gene is incorporated into the bacteria cell which now produces the protein coded for by the gene.
- Bacteria are grown in fomenters. With ideal conditions, the maximum amount of product can be made.
- We can use this method to create useful drugs. For example, we can use it to make the protein insulin.

Now try this

b Match up the two columns.

Trait	Deoxyribonucleic acid
G	A section of DNA that codes for a certain characteristic
Base	Thymine
C	Adenine
DNA	Guanine
A	Cytosine
T	T, A, G and C are examples of these
Gene	A characteristic

Homework

4 Describe how genetically modified bacteria are grown.

5 Draw a flow diagram explaining the stages in inserting an insulin gene into bacteria.

6 Apart from insulin, what other substances might we make using genetically modified bacteria?

ORGANELLES

- **Ribosomes** are **organelles** found within cells.
- Ribosomes produce the specific **proteins** which the **DNA** code for.
- DNA is unravelled.
- **mRNA** is made (a single stranded copy of the DNA).
- mRNA leaves the nucleus and is engulfed by a ribosome in the **cytoplasm**.
- Every three **bases (codon)** code for one specific **amino acid**.
- Many amino acids join together to make a protein.

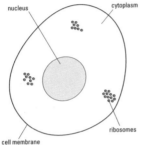

Now try this

c Match up the two columns.

Amino acid	A single-stranded copy of DNA
Ribosome	Three bases make up a _____
Three	Many amino acids join together to make this
Base	Each codon codes for an _____ _____
Codon	The number of bases in each codon
Protein	For example, A, T, G and C
mRNA	This is an organelle that uses RNA to make proteins

Homework

7 What is an organelle?

8 Draw a diagram describing the stages in protein synthesis.

9 Why does DNA have to be transcribed into mRNA in order for proteins to be made?

AEROBIC AND ANAEROBIC

- **Respiration** is a chemical reaction that happens in all cells.
- In the body cells, **oxygen** reacts with **glucose** in a reaction called **aerobic respiration**. **Carbon dioxide** and **energy** are made.
- When exercising hard, the breathing and heart rate must increase so more oxygen can be transported around the body.
- If insufficient oxygen reaches the cells, **anaerobic respiration** occurs. **Lactic acid** and energy are made.
- Anaerobic respiration is less efficient than aerobic.
- After anaerobic respiration, the body must repay the **oxygen debt** accumulated.

Anaerobic means respiration with insufficient oxygen.
AERobic means respiration with sufficient oxygen.

Now try this

d Label the following **aerobic** or **anaerobic**.

Involves plenty of oxygen	_____
Oxygen debt	_____
Produces plenty of energy	_____
Is less efficient	_____
Lactic acid is produced	_____
Occurs in the absence of oxygen	_____
May cause you to get cramp	_____

Homework

10 Write out the word equation and symbol equation for aerobic respiration.

11 Write out the word equation and symbol equation for anaerobic respiration.

12 Explain what we mean by the term 'oxygen debt'.

Divide and develop 1

MITOSIS

- Cells need to divide for growth and repair.
- Body cells are also known as **somatic cells** and contain a full set of **chromosomes (diploid)**.
- Sex cells have half the genetic content of a somatic cell (haploid).
- **Mitosis** is **cell division** where one somatic cell divides to produce two somatic cells, each with a full set (diploid) of chromosomes.

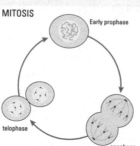

MITOSIS

Early prophase

telophase

anaphase

Now try this

a Fill in the missing words.

Cells must divide in order for the organism to _____ or to _____ damaged parts. _____ is the name for the type of cell division that produces new _____ (body) cells. These cells are _____ as their _____ have been copied.

Homework

1 Name **four** types of specialised somatic cells.
2 Draw a cartoon diagram showing the stages of mitosis.
3 Why must the nucleus double before it can divide?

MEIOSIS

- **Meiosis** is the type of cell division that produces **sex cells**.
- One **diploid** cell divides to produce four **haploid** cells that are not genetically identical.
- Meiosis is important in producing offspring that are not genetically identical to the parents.

TOP TIP

mEiosis produces **Egg** cells. **miTOSis** produces **Toe** cells!

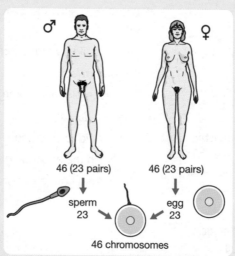

46 (23 pairs)   46 (23 pairs)

sperm 23   egg 23

46 chromosomes

Now try this

b Label the following **mitosis, meiosis** or **both**.

Produces somatic cells   _____

Haploid cells   _____

Produces genetically identical cells   _____

Cell doubles up first   _____

Diploid cells   _____

Produces genetically non-identical cells   _____

Growth   _____

Facilitates variation within species   _____

Produces sperm cells   _____

Reproduction   _____

Homework

4 Explain how meiosis leads to the production of **four** haploid cells.
5 How do meiosis and mitosis differ?
6 Explain why meiosis is an important step in creating variation within a species.

STEM CELLS

- Cells are only able to undergo a certain number of divisions before they die.
- The number of divisions they may undergo is determined by genetics.
- This limit is called the **Hayflick limit**.
- **Stem cells** and cancer cells differ in that they do not have a Hayflick limit.
- A stem cell can **differentiate** into any cell type.
- Some animals are able to regenerate whole limbs or body parts.
- Examples of such animals include spiders, worms and reptiles.
- This is a relatively rare phenomenon as most adult organisms do not have such an availability of stem cells in the body.

Now try this

c Match the key words with the meanings.

i	Genetics	The limit to the number of divisions a cell can undergo
ii	Cancer cell	This type of cell is not yet specialised
iii	Hayflick limit	Mitosis or meiosis
iv	Stem cell	Determine the Hayflick limit
v	Cell division	This type of cell has no Hayflick limit

Homework

7 Find out where we obtain stem cells from.

8 Why is stem cell research banned in some countries?

9 What problems would scientists face should they clone an adult organism from its stem cells?

CONTINUOUS DATA

- Height is a **continuous variable**. This means there is a **range of data**.
- Height in humans is affected by a number of factors such as genes, hormones and nutrition.
- We use a **scatter graph** for continuous data and draw a **line of best fit**. A line of best fit goes through as many of the points as possible.
- Shoe size is **discrete**. This means the measurements fit into certain categories.
- When drawing a graph for discrete data, we tend to use a **bar chart**.

Now try this

d Are the following **discrete (D)** or **continuous (C)**?

Height   _____

Hair colour   _____

Arm length   _____

Shoe size   _____

IQ   _____

Ear shape   _____

Eye colour   _____

Number of leaves _____

Length of leaf _____

Homework

10 What would a graph showing the heights of all the girls in your class look like? Sketch the graph. Repeat for the boys.

11 Explain the differences between the graphs in Homework question **10**.

12 Explain why height is a feature affected by a number of factors.

Divide and develop 2

COMPETITION IN PLANTS

- A number of **factors** limit the growth of plants: light, nutrient availability, temperature, carbon dioxide, oxygen, hormone levels and available space.
- In plant **fertilisation**, the male gamete (**pollen**) lands on a **stigma**. From here, a long tube grows down into the plant **ovule** where it fertilises the female gamete (**ovum**).
- After fertilisation, the petals fall off. Fruit is the swollen **ovary** wall.
- **Seeds** can be transported by: animals and insects, the wind and water.
- The way the seed is **dispersed** affects how far the plant is **distributed**.

 TOP TIP Fields treated with human sewage often have a high number of tomato plants growing.

Now try this

a Fill in the missing words.

The successful growth of a plant seed relies on a number of factors including the availability of _____, _____, _____ and _____. In order to increase the chances of survival, seeds are often _____. Some seeds, are light and are carried through the _____. Other plants form _____, which is eaten by birds and animals. The seeds pass through the _____ tract of the organism and are spread far and wide.

Homework

1 Draw a diagram showing the plant sex organs.
2 Describe the different ways in which seeds can be dispersed.
3 Find out what auxins are and how we use them.

SELECTIVE BREEDING

- **Selective breeding** has been used to create organisms that are more useful to humans.
- For example, we have bred horses that are taller and faster, cows that produce more milk, sheep that produce better wool and crops that produce a greater yield.
- Organisms are chosen according to desired features and are bred together.

Now try this

b Fill in the missing words.

_____ _____ has been used for thousands of years to produce organisms that are more useful for humans. Organisms are chosen for their _____ and are bred. In this way, useful _____ are selected and genes that are not required are _____ out. Examples of organisms that have been selected for human advantage include _____ for their hunting capabilities, _____ for the production of milk, _____ for their speed and _____ for its yield.

Homework

4 If you could add any gene to a plant, what super-plant would you create?
5 Why are organisms such as the fruit fly *Drosophila melanogaster* particularly useful when carrying out genetic experiments?
6 Who was Mendel? Find out about his research.

DOLLY THE CLONE

- A **clone** is an organism genetically identical to its parent.
- A clone is made by taking an egg cell and a normal body cell from an organism.
- The nucleus from the egg cell is removed. This nucleus is **haploid** and has insufficient genetic material to create life.
- The nucleus from a body cell (**diploid**) is then placed into the egg cell.
- The egg is encouraged to divide by being stimulated with electric shocks.
- Once the cell has started to divide, it is replaced into the womb of the mother organism.

Now try this

c Are the following statements **true (T)** or **false (F)**?

Egg cells contain the same number of chromosomes as a liver cell	_____
So far cloning a human is impossible	_____
Egg cells are haploid	_____
A clone is genetically identical to the parent	_____
Cloning does not involve fertilisation	_____
Male cells are not needed when cloning	_____
Cloning: the nucleus from a body cell is placed into an egg cell	_____
A clone has two parents	_____

Homework

7 Why was Dolly the sheep such a scientific breakthrough?

8 Find out why Dolly the sheep died.

9 Draw a cartoon flow diagram describing the steps in cloning.

TREATING GENETIC DISORDERS

- The potential to treat genetic disorders such as cancer and cystic fibrosis is becoming more of a reality.
- Cells affected by the disorder can be 'infected' with genes that do not have the mutation.
- There are many potential ethical dilemmas for such genetic treatment. Some people feel that we have been assigned our own unique genetic code and this should not be tampered with.
- **Genetic therapy** would not necessarily eradicate genetic disorders. Sex cells will still contain the mutation and also spontaneous mutations will still occur.

Now try this

d Circle the genetic disorders.

Cancer

Tonsillitis

Tuberculosis

Cystic fibrosis

Glandular fever

Huntington's chorea

Athlete's foot

Downs syndrome

Malaria

Sickle cell anaemia

Diabetes

Homework

10 Should human cloning be allowed? What are the potential risks?

11 If we were able to treat a genetic disorder with gene therapy, would the disease still be passed on to future generations? Explain your answer.

12 What are the potential ethical dilemmas faced with gene therapy for the relief of genetic disorders?

Energy flow 1

ANIMAL AND PLANT CELLS

- Typical animal cells are made up of a **cell membrane**, **cytoplasm** and a **nucleus**.
- Typical plant cells are made up of a **cell wall**, cell membrane, nucleus, cytoplasm and **chloroplasts**.
- The nucleus contains all the genetic information needed by the organism.
- The cytoplasm is where all chemical reactions take place.
- The cell membrane controls what enters and leaves the cell.

Now try this

a Label the following cell.

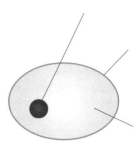

Homework

1 Draw and label a diagram of a typical plant and animal cell. Highlight the similar features and underline the differences.
2 Describe the role of the nucleus, cytoplasm and cell membrane.
3 Research a specialised plant cell and a specialised animal cell. How are these cells different to typical cells?

PHOTOSYNTHESIS

- **Photosynthesis** is a process where **carbon dioxide gas** and **water** are converted into **oxygen** and **glucose**.
- Plants carry out photosynthesis in order to make glucose. They then use this to make energy via **respiration**.
- Photosynthesis needs **light energy**.
- Photosynthesis occurs in the **chloroplasts** in plant cells.
- Chloroplasts are filled with a green pigment called **chlorophyll**.

Now try this

b Match the key words with the meanings.

i	Oxygen	Gas which is a reactant of photosynthesis
ii	Chloroplast	Liquid reactant of photosynthesis
iii	Light energy	Gas produced by photosynthesis and used for respiration
iv	Carbon dioxide	Product of photosynthesis, often stored in the fruit
v	Glucose	Type of energy provided by the Sun, needed for photosynthesis
vi	Chlorophyll	Organelle found in plant cells where photosynthesis occurs
vii	Water	Green pigment

Homework

4 What are the products and reactants of the photosynthesis reaction?
5 Where is the glucose that plants make stored?
6 Why don't root cells have chloroplasts?

USING PLANTS

- Light, carbon dioxide levels and temperature are factors that affect the rate of **photosynthesis**.
- A **limiting factor** of photosynthesis is a factor which prevents the plant carrying out photosynthesis at its **optimum rate**.
- Limiting factors for photosynthesis tend to be either light, water, carbon dioxide or temperature.
- If there is insufficient light, there is not enough energy given for the photosynthetic reaction.
- If there is insufficient water or carbon dioxide, there is not enough reactant for the reaction.
- The optimum temperature for photosynthesis to happen is dependent on the enzymes found within the plant.
- Plants have evolved to compete for these limiting factors.

Now try this

c Are the following statements **true (T)** or **false (F)**?

Glucose is stored in plant roots and fruit _____

Respiration does not occur in plants _____

Photosynthesis is limited by light intensity _____

Photosynthesis occurs in the nucleus of plant cells _____

Photosynthesis does not occur in the roots of plants _____

Chloroplast is a green pigment _____

Homework

7 Explain what we mean by a limiting factor.

8 Explain what the limiting factor is for plants growing on a cold, sunny day.

9 How have plants evolved to receive optimum amounts of light and water?

THE CARBON CYCLE

- The **carbon cycle** describes the cycling of carbon in the form of carbon dioxide.
- Carbon dioxide is taken in by plants for **photosynthesis**. The carbon is locked up in carbon compounds, mainly **glucose** ($C_6H_{12}O_6$). Oxygen is released.
- Living organisms feed on plants. They use the carbon compounds in plants for **respiration**. Carbon dioxide and water are released.
- When plants and animals die, they are broken down by **microorganisms**. Carbon compounds locked up in dead organisms are used for respiration by microorganisms. Carbon dioxide is released.
- **Fossil fuels** are created from dead plant and animal matter, burning them releases carbon dioxide.
- The sea absorbs much carbon dioxide from the air.

Now try this

d Circle the examples where carbon dioxide is released into the atmosphere.

Burning fossil fuels

Respiration

Decomposing

The sea

Production of glucose

Burning wood

Photosynthesis

Homework

10 Draw and annotate the carbon cycle.

11 How has increased levels of carbon dioxide affected the temperature of the Earth?

12 Explain how burning fossil fuels and deforestation has led to an increase in the level of atmospheric carbon dioxide. Write symbol equations to explain.

Energy flow 2

THE NITROGEN CYCLE

- **Nitrogen** is an important **element**, vital for plant growth. It is also a vital component in **amino acids** and therefore **proteins**.
- Special **nitrogen-fixing bacteria** in the soil take nitrogen from the air and convert it into **nitrates** found in the soil.
- Plants take up nitrates in the soil. These are passed into animals when they are eaten.
- When plants and animals die, special bacteria called **decomposers** return the nitrates to the soil.
- Through a series of steps, **nitrifying** and **denitrifying bacteria** convert nitrates back into nitrogen in the air.

Now try this

a Match the words on the left with the definitions on the right.

i Nitrogen-fixing bacteria	Element important in amino acids
ii Decomposers	Make up protein
	Convert nitrogen in the air into nitrates in the soil
iii Nitrogen	Convert nitrates in the soil into nitrogen in the air
iv Amino acids	Break down dead plant and animal matter, returning nitrates to the soil
v Denitrifying bacteria	

Homework

1 Draw the nitrogen cycle. Highlight the role of all bacteria in this cycle.
2 Explain why nitrogen is such an essential element for plants and animals.
3 Why do farmers use fertilisers? What chemicals do they contain?

PROBLEMS WITH FERTILISERS

- Farmers use **fertilisers** to promote plant growth.
- Fertilisers contain **nitrogen**, an element essential in amino acids.
- **Nitrates** are extremely **soluble**. When it rains, they are washed into the water systems.
- The increased use of fertilisers has led to water systems with unusually high levels of nitrates. This results in a process called **eutrophication**.
- High levels of nitrates in lakes cause an increase in the growth of algae. This reduces the amount of light reaching the plants below water. As a result they die.
- The decay of dead plant matter by microbes uses up much of the oxygen in the water and aquatic animal life dies. The dead animal matter makes the situation worse.

Now try this

b Put these steps in order.

- [ ] Aquatic plants decay
- [ ] Murky, dirty water
- [ ] Aquatic animals die
- [ ] Algae are encouraged to grow
- [ ] Aquatic plants start to die
- [ ] Nitrates are leached into lakes
- [ ] Fertilisers are added to the soil
- [ ] Light reaching plants at the bottom of the lake is reduced
- [ ] Oxygen levels in the water are reduced

Homework

4 Draw a flow diagram explaining the process of eutrophication.
5 How is organic farming important in reducing the problem of eutrophication?

TOO MUCH FOOD?

- In some countries, farmers are given incentives to set fields aside. More food than is needed is being produced.
- In many developing countries, there is insufficient food to feed the population. Drought and war add to the problem.
- As the world's human population is increasing, we are placing enormous strains on the Earth to sustain ourselves.
- Energy is lost through the food chain. By restricting energy transfer in our farm animals, we can produce food more efficiently. We can reduce the amount of energy expenditure in farm animals by restricting their movement and controlling the temperature they live in.

Now try this

c Tick the statements that are true.

We can maximise food production by:

Controlling the temperature ☐

Keeping animals indoors ☐

Increasing carbon dioxide levels ☐

Restricting animal movement ☐

Rearing animals outdoors ☐

Feeding animals antibiotics ☐

Protecting animals from predators ☐

Homework

6 Use the Internet to find a graph showing the increasing human population.

7 Describe some of the symptoms of being malnourished.

8 How can growing crops in greenhouses maximise crop yield?

THE GREENHOUSE EFFECT

- Increased levels of **carbon dioxide** and **methane** in the atmosphere cause the **greenhouse effect**.
- The greenhouse effect is a natural effect. Without it, the Earth would be much cooler.
- **Greenhouse gases** occur naturally through the release of gases from volcanoes and methane.
- Burning **fossil fuels** and **deforestation** are human activities which are resulting in the increase in greenhouse gases.
- Greenhouse gases act like a blanket around the Earth.
- Heat from the Sun is trapped, causing the Earth's temperature to rise.
- Rising temperature has caused the melting of the polar ice caps, resulting in a rising sea level.

Now try this

d Tick the activities that lead to an increased greenhouse effect.

Photosynthesis ☐

CFCs ☐

Burning fossil fuels ☐

Using aerosols ☐

Deforestation ☐

Intensive farming ☐

Using fertilisers ☐

Making plastics ☐

The sea absorbing carbon dioxide ☐

Homework

9 Describe some of the consequences of the greenhouse effect.

10 Draw a diagram explaining the greenhouse effect.

11 Find out about the Kyoto Protocol.

Interdependence

POPULATIONS

- A **population** is a group of potentially **interbreeding** species that take up a certain **habitat**.
- Species have evolved to **compete** for factors that increase their chances of survival.
- For example, organisms compete for space. If more space is made available, more organisms will be able to live in that habitat so the population size will increase.
- An increased number of **predators** will result in a population decrease, whereas an increase in **prey** will result in more available food so a larger population will be supported.
- Two groups of factors affect population size: **abiotic** (physical factors such as light and water availability) and **biotic** (competition, predators).

Now try this

a Match the key words with the meanings.

Habitat	A group of interbreeding species in a habitat
Interdependence	A part of the environment in which an organism lives
Predator	The population size of a predator and prey depend on each other
Population	An animal that lives by feeding off other animals

Homework

1 Name **two** biotic and **two** abiotic factors.
2 Describe how the population size of rabbits affects the population size of foxes.

THE ENVIRONMENT

- **Extinction** is a natural evolutionary process but the pressures humans are putting on the Earth is causing an increase in the rate of extinction.
- The human population is growing exponentially. Its effect on the environment is also increasing.
- **Urbanisation** has resulted in the destruction of much of the countryside. When new homes are built, other organisms' habitats are destroyed.
- Burning **fossil fuels**, combined with **deforestation**, has resulted in more carbon dioxide and an increased global temperature.
- **Pollutants** from burning fossil fuels dissolve in rain to make it acidic.
- **Pesticides**, **fertilisers** and **sewage** pollute our water systems.
- **CFCs** deplete the **ozone layer** – a layer around the Earth that protects us from the Sun's harmful rays.

b Fill in the missing words.

Gas released from burning fossil fuels:

_____ _____

Cause eutrophication: _____

Gas reactant for photosynthesis:

_____ _____

Doing this results in acid rain: _____
_____ _____

The human population is growing
_____.

These destroy the ozone layer:
_____

Increased amounts of carbon dioxide result
in _____ _____.

Homework

3 How can wasting electricity ultimately result in global warming?
4 Explain the effect of burning fossil fuels and destroying forests on the carbon cycle.
5 Find out what factors affect the birth and death rates.

PROTECTING OUR ENVIRONMENT

- The increasing size of large towns and cities means a reduction in the countryside.
- Governments are now starting to consider the environment. Schemes are being introduced to encourage homes and businesses to be less wasteful.
- **Biodiversity** is extremely important. **Conservation** efforts are being concentrated in so-called 'biodiversity hotspots'.
- It is important to conserve the environment for a number of reasons:
  - species are **interdependent** – if one species becomes extinct, this will affect other species within the habitat
  - plants are important in absorbing much of the carbon dioxide we release from burning fossil fuels
  - many plants and animals have been vital in providing cures for diseases.

Now try this

c Match the key words with the meanings.

i	Urbanisation	Human intervention to preserve, protect and manage the environment
ii	Biodiversity hotspot	Number and variety of species in an area
iii	Carbon cycle	An area containing a great deal of biodiversity
iv	Biodiversity	Has resulted in the destruction of much of the countryside
v	Conservation	Affected greatly by deforestation

Homework

6 Why is it so important to conserve the environment?

7 Find out what your local council does to protect the environment. What else could it do?

8 Find out about biodiversity hotspots – where are they and why are they so important?

RECYCLING

- In the **recycling** process, materials are reused and made into other items.
- Materials often recycled include glass, paper and many metals.
- Recycling reduces the space needed for waste disposal and means less rubbish is burnt. Both these processes are detrimental to the environment.
- Many local governments provide facilities to recycle household rubbish.
- One problem with recycling is that rubbish must be sorted. Some households find it difficult to separate out their waste.
- Local government recycling facilities are vital in determining whether or not households recycle.

Now try this

d Fill in the missing words.

The process where materials are used again: _____

The process where used materials are taken and used to make new items: _____

Three materials which are often recycled

_____ _____ _____

This method of waste disposal uses up a lot of space: _____ _____

This method of waste disposal releases many toxic gases into the atmosphere: _____

Homework

9 Explain why reusing is more efficient than recycling.

10 What else could be done to encourage people to recycle more?

Carbon compounds

HYDROCARBONS

- Living things contain millions of **carbon** compounds.
- Carbon atoms make four bonds and can form chains.
- The simplest are **hydrocarbons**.
- **Alkanes** like methane, ethane, propane and butane contain single **covalent bonds** and are useful fuels. Their **general formula** is C_nH_{2n+2}.
- **Alkenes** like ethene and propene have C=C **double bonds**, which makes them reactive. Their general formula is C_nH_{2n}.

ethane C_2H_6

ethene C_2H_4

Now try this

a Match each hydrocarbon to its name.

C_2H_6	methane
CH_4	propane
C_3H_6	butane
C_3H_8	ethane
C_4H_{10}	propene

 Count the bonds around each carbon – there must be four.

Homework

1. Draw these hydrocarbons and say which are alkanes: C_6H_{14}, C_5H_{10}, C_7H_{16}, C_8H_{16}.
2. Suppose butane split in two. Draw **two** hydrocarbons made from butane's atoms.
3. Find out how isomers differ and draw the **two** isomers of butane.

CRACKING

- **Cracking** breaks **alkanes** into more useful fuel molecules and **alkenes** for **polymer** formation.
- **Bromine water** turns colourless with alkenes, because it adds to C=C double bonds.
- Water adds to alkenes to make **ethanol**.
- The double bonds make alkenes **unsaturated**.
- Alkanes have no double bonds, so they are **saturated**. They do not react with bromine.

 Alkanes crack when they are heated – not when they burn.

Now try this

b Match each word to its definition.

i	Alkane	Contains a C=C bond
ii	Saturated	C_2H_5OH
iii	Cracking	Saturated hydrocarbon
iv	Alkene	Unsaturated hydrocarbon
v	Hydrocarbon	Decomposition
vi	Unsaturated	Contains no C=C bonds
vii	Ethanol	Compound of H and C

Homework

4. Find out what ethene and propene are used for.
5. Show what happens to the molecules when ethene and water react.
6. Research the industrial cracking of hydrocarbons and explain why it is useful.

POLYMERS

- **Addition polymers** form when C=C bonds break and **unsaturated monomers** link up. For example, propene makes polypropene.
- **Thermosetting polymers** stay hard because their chains are held in place by **cross-links**.
- **Thermoplastics** soften and melt because their chains slide over each other.

 TOP TIP The C=C bonds change to C–C when polymers form.

Now try this

c Match each monomer to its polymer.

Styrene	Polypropene
Ethene	
C_4H_8	PVC
C_2H_4	Polyethene
C_3H_6	
Vinyl chloride	Polybutene
Chloroethene	Polystyrene

Homework

7 Prepare a presentation to explain how monomers become polymers.

8 Find the names and uses of **three** thermosetting and **three** thermosoftening plastics.

9 Find out how the chains differ in high density and low density polyethene.

PLASTICS

- We depend on oil but it is **non-renewable**.
- We will need new fuels and raw materials.
- **Polymer** properties depend on the **monomer** used, the reaction conditions, whether it cross-links, and the presence of additives like **plasticisers** and **preservatives**.
- Cross-links make plastics more rigid.
- Plasticisers make them more flexible.
- Preservatives stop them becoming brittle.
- **Plastics** are **non-biodegradable**. Some polymers release toxic products when burned.
- Plastics are coded to aid **recycling**.

Now try this

d Tick the statements that are true.

i Cross-links make plastics easy to mould ☐

ii Preservatives make more flexible ☐

iii Plasticisers are used in PVC window frames ☐

iv Recycling codes can be used to identify plastics ☐

v Plastics break down in landfill sites ☐

vi Most polymers are made from oil ☐

 TOP TIP Additives can make plastics break down in sunlight.

Homework

10 Find out what the numbers 1–6 mean on plastic recycling codes.

11 Find out what plasticisers are, and explain what they do to PVC.

12 Prepare a presentation on ways of changing a plastic's properties.

Making changes

FATS AND OILS

- Vegetable **oils** contain chains of **carbon** atoms.
- Those with one C=C bond are **monounsaturated**.
- Those with several C=C bonds are **polyunsaturated**.
- Extra C=C bonds make oils more runny (less **viscous**)
- Vegetable oils are reacted with H_2 to reduce the number of C=C bonds and make **fats** for margarine.

 TOP TIP Adding hydrogen to a C=C bond is an addition reaction.

Now try this

a Fats A and B each contain 16 carbons. A is saturated, B is polyunsaturated. Match each description below to **A** or **B**.

No double bonds	_____
Highest melting point	_____
Does not decolourise Br_2	_____
Best for your heart	_____
Contains C=C bonds	_____
More viscous	_____
Most runny	_____

Homework

1 Draw a table to show the types of fat in **four** brands of margarine.

2 Prepare a presentation to show how oils react with hydrogen.

3 Research the health problems blamed on saturated fats.

NEW PRODUCTS

- Similar chemicals react the same way, for example all alkenes react with bromine; so a reaction's products are predictable.
- New chemicals are checked for **toxicity**.
- **Drug development** involves four steps:
  - Find a **target protein** for the drug to act on.
  - Use **staged synthesis** to make mixtures containing thousands of new chemicals. For example, if two compounds react with two others, four products go through to the next stage.
  - Test the chemicals to find one that affects the target and has no serious side-effects.
  - Carry out **clinical trials** on humans.
- Very few drugs get through all the tests, so developing them costs millions.

 TOP TIP Patents protect a company's investment by stopping other companies making copies for 20 years.

Now try this

b Add the missing words.

two	**target**	**mixtures**
clinical	**four**	**compounds**

Drug companies make more than 100 000 new _____ every year. Staged synthesis methods let them make _____ of new compounds quickly. Foir example, if A reacts with a mixture of B and C, _____ products form. If D and E react with F and G, there are _____ possible products. Robotic techniques are used to test the products against _____ proteins. Promising drugs undergo _____ testing.

Homework

4 Many existing drugs come from plants. Find some examples.

5 Find out what reactions are used when aspirin is synthesised.

6 Find out how penicillin was discovered.

EQUATIONS

- The **relative formula mass** is the sum of the masses of the atoms.

    $CO_2 = 12 + 16 + 16 = 44$

    $CaCO_3 = 40 + 12 + 16 + 16 + 16 = 100$

- Formulae cannot change but **equations** must be **balanced**.

    $2H_2O$ means $H_2O + H_2O$

- Equations can be used to calculate how much product will form.

    $CaCO_3 \rightarrow CaO + CO_2$

    100 g $CaCO_3$ make 44 g CO_2

    1 g $CaCO_3$ makes 0.44 g CO_2 and so on.

$$2H_2 + O_2 \rightarrow 2H_2O$$

Two molecules of water must be made to use up both oxygen atoms.

Multiply everything in brackets by the number outside.

Now try this

c Calculate the formula masses, and balance the equations.

CaO _____

C_2H_6 _____

C_2H_5OH _____

$Ca(OH)_2$ _____

$H_2 + I_2 \rightarrow$ _____ HI

_____ Mg + $O_2 \rightarrow$ _____ MgO

_____ Na + $Cl_2 \rightarrow$ _____ NaCl

Mg + _____ HCl $\rightarrow MgCl_2 + H_2$

Mg + _____ $HNO_3 \rightarrow Mg(NO_3)_2 + H_2$

$C_2H_4 +$ _____ $O_2 \rightarrow$ _____ $CO_2 +$ _____ H_2O

$C_3H_8 +$ _____ $O_2 \rightarrow$ _____ $CO_2 +$ _____ H_2O

Homework

7 Prepare a presentation to show Year 7 students how to find a compound's mass.

8 Put these gases in order of mass: CH_4, CO_2, Cl_2, H_2, N_2, SO_2, NH_3, NO_2.

9 Calculate the CO_2 made when these burn: 160 g of CH_4, 7 g of C_2H_4, 96 g of C_3H_8.

CALCULATIONS

- The **atom economy** of a reaction shows the percentage of reactant atoms converted to product.

- Atom economy = $\dfrac{\text{useful product (g)}}{\text{total (g)}} \times 100\%$

- It depends on the **yields** of the reactions.

- Yield = $\dfrac{\text{actual product (g)}}{\text{theoretical product (g)}} \times 100\%$

- High atom economies prevent waste.

- Reacting masses can be used to find the **empirical formulae** of compounds. For example:

    S has a mass of 32, O has a mass of 16. If 32 g of S react with 32 g of O, one atom of S reacts with two atoms of O.

    Formula = SO_2

High atom economies make reactions more profitable.

Now try this

d Calculate each formula.

48 g of Mg with 32 g of O _____

20 g of Ca with 8 g of O _____

24 g of C with 64 g of O _____

24 g of C with 8 g of H _____

8 g of S with 8 g of O _____

28 g of N with 6 g of H _____

24 g of C with 6 g of H _____

8 g of H with 64 g of O _____

Homework

10 48 g of Mg make 32 g of MgO. What is the percentage yield?

11 What is the atom efficiency if 80 g out of 200 g are the right product?

12 Research the atom efficiency of the reactions used to make ibuprofen.

Elements

METALS

- **Metals** are good **conductors** because they have **delocalised electrons**.
- Strong bonds between the atoms give them high melting and boiling points.
- They are hard but **malleable**, because atoms can slide into new positions.
- **Alloys** are mixtures of metals.
- Alloying with bigger or smaller atoms disrupts crystals and changes properties.

Now try this

a Match these beginnings and ends.

Strong bonds cause	are mixtures of metals
	high tensile strength
Delocalised electrons	make metals malleable
	cause conductivity
Alloys	high melting points
	are negatively charged
Flexible bonds	have useful properties

TOP TIP Alloys are often more useful than pure metals.

Homework

1 Find **three** different images of atoms in a metal.
2 Find the name, composition and use of **five** different alloys.
3 Prepare a presentation to show why metals have the properties they do.

ATOMS

- Each **element** has a different **atomic number**.
- It shows the number of **protons**.
- Positive protons and neutral **neutrons** both have a **relative mass** of 1.
- Their total gives the atom's **mass number**.
- Negative **electrons** have negligible mass.
- **Isotopes** have different numbers of neutrons, so they have different mass numbers.
- The **relative atomic mass** of an element is the average mass of its atoms.

Average mass of chlorine atoms = 35.5.

mass = 35   mass = 35   mass = 37

Now try this

b Pick the particle(s) that decide:

Atomic number	Electrons
Mass number	
The element	Protons
Relative atomic mass	
	Neutrons
Positive charge in nucleus	
Position in periodic table	Isotopes

TOP TIP The relative atomic mass of Cl is 35.5: three quarters have mass 35 and a quarter have mass 37.

Homework

4 How many protons, neutrons and electrons are there in H, B, Na, Al, K and Ca?
5 How did Mendeleev construct the first periodic table?
6 Design a test to cover everything in the atoms panel.

ELECTRONS

- Protons and neutrons stay in the nucleus.
- **Electrons** orbit in **shells**.
- The periodic table reflects these shells.
- The first shell holds two like the first row.
- The second shell holds eight like the second row.
- Electronic configurations show the electrons in each shell. For example, C (2, 4) has two in the first shell, four in the second.
- The number of electrons in the outer shell decides how an element reacts.
- **Mendeleev** showed creative insight by leaving gaps for unknown elements, so similar elements are in the same group, for example group 1 contains reactive metals.

 Group number is equal to electrons in the outer shell, and this determines an element's properties.

Now try this

c Write electron configurations for:

B _____

N _____

Ne _____

Mg _____

Cl _____

Ca _____

Al _____

P _____

Homework

7 Compile a 10-question quiz on the periodic table. Supply answers on the back.

8 Find images of **five** elements with less than 20 protons and draw their atoms.

9 Write down everything you know about K, Cl and Ne from the periodic table.

IONS

- Atoms are **neutral** because they have equal **protons** (+) and **electrons** (–).
- In reactions, metals lose electrons.
- Their atoms become **positive ions**, with more protons than electrons. For example:
  $Na_{(2, 8, 1)} - 1$ electron $\rightarrow Na^+_{(2, 8)}$
- The electrons are given to non-metals, so non-metals make **negative ions**. For example:
  $Cl_{(2, 8, 7)} + 1$ electron $\rightarrow Cl^-_{(2, 8, 8)}$
- The Na and Cl make an **ionic bond**.
- Enough electrons transfer to give both metal and non-metal a full outer shell. For example:
  $Mg_{(2, 8, 2)} - 2$ electrons $\rightarrow Mg^{2+}_{(2, 8)}$
  $Cl_{(2, 8, 7)} + 1$ electron $\rightarrow Cl^-_{(2, 8, 8)}$

  So two Cl take one electron each.
  $2Cl + 2$ electrons $\rightarrow 2Cl^-$

 Electrons transfer from metal to non-metal when ionic bonds form.

Now try this

d Finish these equations.

$Mg_{(2, 8, 2)} - 2$ electrons $\rightarrow$ _____

$Cl_{(2, 8, 7)} + 1$ electron $\rightarrow$ _____

$Na_{(2, 8, 1)}$ _____

$Al_{(2, 8, 3)}$ _____

$O_{(2, 6)}$ _____

$K_{(2, 8, 8, 1)}$ _____

Ca _____

F _____

S _____

Li _____

Homework

10 Draw electron diagrams to show how MgO, BeF_2 and Na_2O form.

11 Find images of giant ionic structures and explain why ions form crystals.

12 Research what happens to the ions when compounds like NaCl dissolve.

Explaining properties

GIANT IONIC STRUCTURES

- **Ionic compounds** have giant structures.
- **Ions** have charge, but compounds are neutral.
- The + and – charges must cancel out, for example, $2Na^+$ with $1O^{2-}$ makes Na_2O.
- Ions like OH^- and NO_3^- may need brackets, for example, Mg^{2+} with $2OH^-$ makes $Mg(OH)_2$.
- The + and – ions are strongly attracted, so melting and boiling points are high.
- Molten, or dissolved, ions move around, so they can **conduct** electricity.

 Use + and – charges to work out formulae. Then cancel them out.

Now try this

a Predict the formula of each compound.

Potassium (K^+) oxide (O^{2-}) = _____

Calcium (Ca^{2+}) chloride (Cl^-) = _____

Sodium (Na^+) iodide (I^-) = _____

Calcium (Ca^{2+}) hydroxide (OH^-) = _____

Calcium (Ca^{2+}) oxide (O^{2-}) = _____

Potassium (K^+) sulphide (S^{2-}) = _____

Calcium (Ca^{2+}) nitrate (NO_3^-) = _____

Sodium (Na^+) sulphate (SO_4^{2-}) = _____

Magnesium (Mg^{2+}) iodide (I^-) = _____

Lithium (Li^+) oxide (O^{2-}) = _____

Homework

1. It is possible to predict the formula of any ionic compound. Explain how.
2. Explain how the forces between ions decide a compound's properties.
3. Produce a leaflet to show Na and how things will change if it reacts with Cl.

ELECTRONS

- The **noble gases** are in **group 8**.
- Their outer **electron shells** are full.
- Full shells are very stable, so they are completely unreactive.
- Other non-metals must gain one, two or three **electrons** to get full outer shells.
- Metals have three outer electrons or less, and lose them all when they react.
- When electrons are transferred, the **ions** formed are stable and unreactive because they have full outer shells.
- **Group 4** react, but do not form ions.

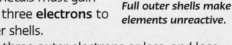

Helium

Neon

Full outer shells make elements unreactive.

 When elements make ions they get the electron configuration of the nearest noble gas.

Now try this

b Decide if each element is in group 1, 7 or 8.

Forms 1+ ions ☐

An unreactive gas ☐

Has seven outer electrons ☐

Gains one electron when it reacts ☐

A reactive non-metal ☐

Forms 1– ions ☐

Atoms have full outer shells ☐

Homework

4. Label a periodic table to show the ions made by groups 1–3 and 5–7.
5. Find out how and when helium was discovered.
6. Research the discovery of potassium.

TRENDS IN REACTIVITY

- Atoms get bigger going down each group, by adding more electron shells, so their physical properties and reactivity change.
- Metals react by losing electrons.
- Bigger atoms lose electrons more easily, because they are further from the nucleus, so the **order of reactivity** in group 1 is K, Na, Li.
- Non-metals react by gaining electrons.
- Smaller atoms gain electrons more easily as their outer shell is closer to the nucleus. The order of reactivity in group 7 is F, Cl, Br, I.
- A more reactive halogen will **displace** a less reactive one from its compounds. For example:
$Cl_2 + 2KI \rightarrow 2KCl + I_2$.
- Group 7 molecules are coloured but their ions are colourless.

Now try this

c Finish these equations.

sodium + water → sodium hydroxide + _____

$2Na + 2H_2O \rightarrow 2NaOH +$ _____

potassium + water → _____

$2K + 2H_2O \rightarrow$ _____

chlorine + potassium bromide
→ potassium chloride + _____

$Cl_2 + 2KBr \rightarrow 2KCl +$ _____

bromine + potassium iodide → _____

$Br_2 + 2KI \rightarrow$ _____

 Large metal atoms and small non-metal atoms are most reactive.

Homework

7 Based on what you know about sodium and potassium, predict the properties of caesium.

8 What patterns are there in the colours, states and melting points of group 7?

9 Find out how bromine is extracted from sea water.

ELECTROLYSIS

- **Electrolysis decomposes** compounds by turning their **ions** back into atoms.
- The ions are melted or dissolved so that they are free to move.
- Then + and – electrodes are connected.
- **Metal cations** (+) move to the negative electrode and get turned into atoms. For example:
$Pb^{2+} + 2$ electrons $\rightarrow Pb$.
- **Non-metal anions** (–) are attracted to the **positive electrode** where electrons are removed. For example:
$2Br^- - 2$ electrons $\rightarrow Br_2$.

 During electrolysis of solutions, hydrogen may be given off instead of a metal. It comes from the water.

Now try this

d Fill in the blanks.

During electrolysis the _____ in molten lead bromide are attracted to the _____. Opposites attract so the Pb^{2+} ions move to the _____ where they gain _____ and become atoms. The Br^- ions _____ to the _____ where they _____ an electron and pair up to make _____. This carries on until all the lead bromide has been _____.

Homework

10 Show what happens to the ions during the electrolysis of $PbBr_2$, and solutions of $CuCl_2$ and HCl.

11 Find out what products form when a solution of NaCl is electrolysed.

12 Prepare a presentation to explain electrolysis to a Year 9 class.

Molecules and metals

MOLECULES

- All gases, water, and some solids are made from small molecules like CO_2.
- They form when non-metals share **electrons** to get full outer **shells**.
- Number of **bonds** formed = (8 − group number). For example:
  C=4, N=3, O=2, Cl=1 and H=1.
- C and O form double and single bonds, for example, O=O in O_2 and O=C=O in CO_2.
- N_2 has a triple bond but NH_3 has three singles.
- Each single **covalent bond** contains one electron from each atom.
- Simulation software can show molecular shape more clearly than 2-D diagrams.

 Always draw the atom with most bonds first – it is usually in the middle of the molecule.

Now try this

a Draw the bonds in these molecules.

H H

N N

Cl Cl

O O

H Cl

H O H

O C O

Homework

1 Draw the bonds in NH_3, CH_4, C_2H_6, C_2H_4, C_2H_5OH and CH_3COOH.
2 Find images of the atoms in diamond, graphite, a fullerene and a nanotube.
3 Find the formula of glucose and draw its bonds.

PURE CARBON

- **Diamond** and **graphite** are ordered networks of atoms called **giant covalent structures**.
- The strong **covalent bonds** give them high boiling points and make them **insoluble**.
- In diamond, each carbon forms four bonds to make a strong 3-D network.
- In graphite, each carbon bonds to three others in a layer.
- The layers are loosely attracted to each other.
- Each atom has a **delocalised electron**, so graphite **conducts**.
- **Fullerenes** are hollow balls similar to graphite.
- In **nanotubes** the carbon sheets roll into tubes, which have a very high tensile strength.

 Carbon nanotubes are stronger than steel.

Now try this

b Does each description match **diamond (D)** or **graphite (G)**?

Soft _____

Transparent _____

Used in pencils _____

Has delocalised electrons _____

Used in drill bits _____

A lubricant _____

Insulator _____

Hard _____

Abrasive _____

Opaque _____

Conducts _____

Homework

4 Explain why graphite is a good conductor and lubricant but diamond is not.
5 Find out what nanotubes could be used for.
6 Find out what role chance played in the discovery of fullerenes.

MOLECULAR PROPERTIES

- The **electrons** in **covalent bonds** can be shown as dots and crosses.
- Each bond has an electron from each atom.
- The bonds are strong but the forces between small molecules like Cl_2 are weak, so their melting and boiling points are low.
- Stronger forces exist between bigger molecules.
- Small molecules like Cl_2 have no **delocalised electrons** or **ions**, so they never conduct electricity.

Now try this

c Draw dots and crosses to show the outer electrons of the atoms in these molecules.

H_2 H   H

H_2O H   O   H

O_2 O   O

N_2 N   N

 TOP TIP A high melting point shows there are strong forces between the molecules.

Homework

7 Use group 7 to explain how size affects the forces between molecules.

8 Find out how SiO_2 is bonded, and explain why it is so different from CO_2.

9 Compare the molecules with the lowest and highest melting points you can find.

METALS

- **Metals** are giant structures of atoms.
- Each atom contributes **delocalised electrons** to the **crystal structure**.
- When a voltage is applied, these move and create an electric current.
- Unlike ionic compounds, metals do not decompose when they **conduct**.
- Free electrons bond atoms together, but allow them to slide; so metals change shape without breaking.

 TOP TIP Most metals are dense materials with high melting points because metallic bonding is very strong.

Now try this

d Decide if the following are **metallic (Me)**, **molecular (Mo)** or **giant covalent structures (G)**.

Gas   _____

Bendy conductor   _____

Liquid insulator   _____

Hard non-conductor   _____

Liquid conductor   _____

Lubricant that conducts   _____

Easily melted solid   _____

Hard conductor   _____

Homework

10 Find **two** images of metal atoms and say which best explains their properties.

11 Suggest reasons to account for some metals conducting better than others.

12 Compare the conductivities of copper and different forms of carbon.

Using chemistry

TESTING PHARMACEUTICALS

- Chemicals are used as **antibiotics** and to counteract problems that cause ill health.
- New chemical-based therapies have to be tested for safety in three stages:
  - Stage 1: potential drugs are **tested** on isolated human cells grown in a laboratory.
  - Stage 2: they are then tested on whole animals to check for **toxicity**.
  - Stage 3: drugs undergo **clinical trials** on volunteers, then on a few patients and finally on large numbers of patients.
- Most trials are **double blind** to make sure the testers are unbiased. Neither patient nor doctor knows who receives the new drug and who gets an existing drug or a **placebo** (fake drug).

Now try this

a Add the missing words.

> lab   volunteers   placebo   therapies
> clinical   animals   existing

New chemical-based _____ are not allowed on sale until they have undergone _____ trials. First they are tested in the _____, then on _____. If a drug appears safe it is tested on healthy _____. A new drug must work better than any _____ drug or a _____.

 Drugs are compared with placebos because even fake drugs make you feel better for a while.

Homework

1 Research the placebo effect.
2 Find out what went wrong when TGN1412 went for clinical trials.
3 The first clinical trials are never done on young women. Why is this?

HOMEOPATHIC MEDICINES

- **Homeopathy** treats patients' **immune systems** with extremely dilute solutions of toxic chemicals.
- Most scientists argue that homeopathic remedies are too dilute to work because no molecules of the original chemical are left – just water.
- Some claim that links between water molecules retain a 'memory' of the original chemical.
- Some **double-blind trials** have shown that homeopathy has no effect, but critics argue that the methods used were to blame.
- Scientists find homeopathy hard to accept because there are no known **metabolic processes** that 'modified' water can affect.

 Before science journals accept new results for publication they are peer-reviewed to check that proper controls were used.

Now try this

b Label the following statements A–D to show which are most likely to convince scientists that homeopathy works against hay fever (A = most likely).

_____ 38 per cent of the 2000 patients in a double-blind trial were cured.

_____ A peer-reviewed article shows pure water and homeopathic solutions have different properties.

_____ You tried it and you have never felt so well.

_____ Horizon filmed some successful laboratory tests.

Homework

4 Write arguments for and against using a homeopathic hay fever remedy.
5 Research **two** other forms of 'alternative' medicine and how they might work.
6 Plan a test to convince sceptics that your new remedy really does work.

MAKING NEW SUBSTANCES

- Chemical reactions make new substances.
- Reactions happen when particles collide.
- The rate of the reaction increases when the **frequency** or **energy of collisions** increases.
- During reactions, **bonds** break in the original molecules and new ones form in the products. For example:
  $$H-H + Cl-Cl \rightarrow H-Cl + H-Cl.$$
- Some atoms are joined by double bonds. For example:
  $$C + O=O \rightarrow O=C=O$$
- The bonds are not usually shown in equations:
  $$H_{2(g)} + Cl_{2(g)} \rightarrow 2HCl_{(aq)}$$

TOP TIP Only the most energetic collisions supply enough energy to break bonds.

Now try this

c Write equations for the following.

i Hydrogen reacting with iodine

_____

ii Hydrogen reacting with oxygen

_____

iii Iron sulphide formation

_____

iv Ammonia (NH_3) formation

_____

Homework

7 Use sodium chloride formation to show that elements change when they react.

8 Write equations for elements from Na to Cl reacting with oxygen and bromine.

9 Draw diagrams to show the bonds broken and formed when methane (CH_4) burns.

ENERGY TRANSFER

- Bond-breaking is **endothermic** – heat energy is taken in.
- Bond-making is **exothermic** – heat energy is given out.
- If the products have stronger bonds, energy is released, the temperature rises and the reaction is exothermic. For example: $2H_2 + O_2 \rightarrow H_2O$ (and any other fuel burning).
- Exothermic reactions are also used in hot packs.
- If the products have weaker bonds, energy is taken in – the reaction is endothermic, for example, baking cakes or clay.
- Some endothermic changes take heat from their surroundings and cause a temperature drop. For example:
  $$NH_4NO_{3(s)} + H_2O_{(l)} \rightarrow NH_4NO_{3(aq)}.$$
- Endothermic reactions are used in cold packs.

TOP TIP Most exothermic reactions need an energy input to start them.

Now try this

d Indicate whether each reaction is **exothermic (Ex)** or **endothermic (En)**?

A burning fuel	_____
Bond formation	_____
Used in fireworks	_____
Used in cold packs	_____
The reactants explode	_____
Needs continual heating	_____
Product has stronger bonds	_____
Used in self-heating cans	_____
The reactants lose energy	_____
Causes a temperature drop	_____
The reactants gain energy	_____

Homework

10 Classify photosynthesis and respiration as exothermic or endothermic. Give reasons.

11 Prepare a presentation to explain what bonds have to do with energy changes.

12 Research the reactions that have been used to make self-heating food cans.

Reaction rates

INVESTIGATING RATES

- Higher **concentrations** make **collisions** more frequent, so **reaction rates** go up. For example, sodium thiosulphate gets cloudy faster with more concentrated hydrochloric acid.
- Increasing the **surface area**, by cutting solids into small pieces, also makes collisions more frequent. For example, smaller marble chips react faster with acid. More atoms are exposed on the surface of the solid.
- At higher **temperatures** reactants have more energy. They collide more frequently, and more violently. For example, sodium thiosulphate gets cloudy faster if it is heated before acid is added.

TOP TIP The rate shows the reactant used, or product made, in a certain time.

Now try this

a Indicate whether the following will make the reaction between magnesium and acid **faster (F)**, **slower (S)** or **neither (N)**.

Adding water to the acid	_____
Cutting the magnesium up	_____
Lowering the temperature	_____
Letting the magnesium float	_____
Using more concentrated acid	_____
Reusing the same acid	_____
Using more acid	_____

Homework

1. Make a list of ways you can measure the rates of reactions.
2. Explain these cooking times: crisps 30 seconds, chips 3 minutes, boiled potatoes 20 minutes.
3. Explain why global warming and acid rain make limestone wear away faster.

CATALYSTS

- **Catalysts** increase **reaction rates** by reducing the energy a collision needs to supply.
- They allow reactions to take place at low temperatures, which reduces costs.
- **Enzymes** are **biological catalysts**. Life would be impossible without them.
- Graphs are easier to produce if datalogging equipment is used to monitor the reaction.
- Pressure sensors can monitor gas production.
- Light sensors detect cloudiness/colour change.
- Temperature sensors detect energy transfer.
- pH sensors detect acid production.
- Mass sensors can detect the loss of a gas.
- The values can be shown on a graph.
- The steeper the graph, the faster the reaction.

Now try this

b Choose an appropriate sensor for each reaction.

A solution that turns cloudy	_____
A very exothermic reaction	_____
Acid reacting with alkali	_____
It makes a heavy gas	_____
It makes a very light gas	_____
It forms a dark blue product	_____
$Mg_{(s)} + 2HCl_{(aq)} \rightarrow MgCl_{2(aq)} + H_{2(g)}$	_____

TOP TIP Catalysts are expensive but they are not used up. They can be used again.

Homework

4. Research **two** industrial uses of catalysts.
5. Design a web page to market a new discovery – a universal catalyst.
6. Find out how fast reactions can get and what chemists use to measure them.

REVERSIBLE REACTIONS

- Some chemical reactions are **reversible**.
- They reach **equilibrium** in closed containers.
- The forward and backward reactions go at the same rate so there is no overall change. For example:

$$C_2H_{4(g)} + H_2O_{(g)} \rightleftharpoons C_2H_5OH_{(g)}$$

- Once equilibrium is reached, ethanol breaks down as fast as it forms – but if it is removed, the equilibrium shifts to restore the **percentage yield**.
- The yield (% product) is increased by:
  - raising the concentration of a reactant
  - lower temperatures (if exothermic)
  - higher temperatures (if endothermic)
  - higher pressures (if gases combine).

 Catalysts don't affect position of equilibrium (yield) because they have the same effect on the forward and backward reactions.

Now try this

c $A_{(g)} + B_{(g)} \rightleftharpoons C_{(g)}$
The forward reaction is exothermic.

Say if the following will cause more or less of C to be made or neither.

Raising the pressure _____

Raising the temperature _____

Continually removing C _____

Adding more A or B _____

Adding a catalyst _____

$\rightleftharpoons$ *This symbol shows that the reactants are not all converted to products.*

Homework

7 Find the temperature and the catalyst used for **two** reversible reactions.

8 Find out what Le Chatelier said about equilibrium reactions.

9 Find out why weak acids are not as corrosive as strong ones.

AMMONIA

- **Ammonia** is made using a reversible reaction:

$$N_{2(g)} + 3H_{2(g)} \rightleftharpoons 2NH_{3(g)}$$

- The conditions used are a compromise: 450 °C, 200 atmospheres, **iron catalyst**.
- Higher pressures would increase the **equilibrium yield** but be dangerous and expensive.
- Lower temperatures would increase the equilibrium yield but make the reaction too slow.
- Most ammonia is used to make **fertilisers**.
- Most farms use artificial fertilisers.
- Organic farms have lower crop yields but may be more sustainable.
- Conventional farming takes a lot of energy, damages the soil, reduces biodiversity and causes pollution.

 Ammonia is neutralised with acid to make fertilisers like ammonium nitrate – NH_4NO_3.

Now try this

d Match the molecules to the descriptions.

N_2	Nitrogen
	Ammonium
NH_3	Molecule
	Found in air
	Ion
NH_4^+	Nitrate
	In fertiliser
NO_3^-	Ammonia

Homework

10 Find out what causes 'blue baby' syndrome and 'eutrophication'.

11 Find out how ammonia is converted to fertilisers like ammonium nitrate.

12 Find out who invented the Haber process. What were the consequences?

Understanding motion

SPEED, VELOCITY, ACCELERATION

- A **scalar quantity** has only magnitude (size). Examples include **speed**, time and distance.
- A **vector quantity** has magnitude (size) and direction. Examples include **displacement**, **velocity**, **acceleration** and **force**.
- speed = $\dfrac{\text{distance}}{\text{time}}$ (Units: m/s, km/h, etc.)
- Displacement is distance in a given direction (for example, 10 km due east).
- Velocity may be defined as the speed in a certain direction.
- average velocity = $\dfrac{\text{displacement}}{\text{time}}$ (Units: m/s, km/h, etc.)
- Velocity of an object can change if its direction or its speed changes.
- Acceleration is equal to the rate of change of velocity.
- acceleration = $\dfrac{\text{change in velocity}}{\text{time}}$ or $a = \dfrac{v - u}{t}$
- Acceleration is measured in m/s².
- The **gradient** from a displacement–time graph is equal to velocity.
- A negative velocity means that the object is moving in the opposite direction.
- The gradient from a velocity–time graph is equal to acceleration.
- A negative acceleration means that the object is **decelerating** (slowing down).

Now try this

a Circle the correct answers.

i What type of quantity is speed?

vector   scalar

ii What type of quantity are both velocity and acceleration?

vector   scalar

iii What has the unit m/s²?

speed   acceleration

iv What is the gradient from a velocity–time graph equal to?

velocity   acceleration

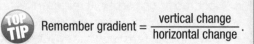

TOP TIP Remember gradient = $\dfrac{\text{vertical change}}{\text{horizontal change}}$.

Homework

1 Make a list of all the quantities in this section and their associated units.

2 Using sketches, describe the information that you can get from displacement–time and velocity–time graphs.

3 Use the Internet to search for an animal or insect with the largest acceleration. Make a list of your websites for the class.

FORCE AND MOTION

- An object remains at rest or travels at constant **velocity** when acted upon by a **resultant force** of zero.
- An object has **acceleration** when the resultant force is not zero.
- For an object of a given mass, doubling the force will double the acceleration.
- For a given force, doubling the mass of an object will halve its acceleration.
- resultant force = mass × acceleration
- Force is measured in **newtons** (N).
- A force of 1 N acting on an object of mass 1 kg causes an acceleration of 1 m/s^2.
- When two objects collide ('interact'), the force they exert on each other is the same but in opposite directions – 'action' = 'reaction'.
- **momentum** = mass × velocity (Unit: kgm/s^2)
- Momentum of an object is a **vector quantity**.

Now try this

b Are the following statements **true** or **false**?

i Acceleration is measured in 'metres per second'. _____

ii The resultant force on an object travelling at constant velocity is zero. _____

iii acceleration = $\dfrac{\text{resultant force}}{\text{mass}}$ _____

iv An object is pulled in opposite directions with forces of 3 N and 4 N. The resultant force is 7 N. _____

Homework

4 List all the equations from this section.

5 Why do less massive cars have better acceleration than more massive cars?

6 Make a list of situations where 'Action' = 'Reaction' would be applicable.

MASS OR WEIGHT?

- The **mass** of an object, in kilograms, remains the same.
- The **weight** of an object is the **gravitational force** acting on the object due to the Earth.
- Weight is a **force** – it is measured in **newtons** (N).
- The weight of an object depends on where it is. The weight of an object on the Moon is less than its weight on the Earth.
- weight = mass × gravitational acceleration ($W = mg$, $g = 10$ m/s^2 on the Earth's surface.)

Now try this

c Find the following key words in this word search.

mass   weight   force   newtons   Earth

W	N	A	H	T	R	A	E
T	G	E	K	A	V	A	L
D	B	A	W	A	S	H	T
S	D	U	E	T	F	D	E
S	S	T	I	Q	O	X	E
A	G	E	G	F	R	N	N
M	P	E	H	C	C	Y	S
G	L	N	T	V	E	U	M

Homework

7 Write bullet points for the term 'weight'.

8 Use the Internet to find the surface gravitational acceleration for some of the planets in our Solar System. Present the information in a table.

Falling and collisions

FALLING DOWN

- A falling object in air has two **forces** acting on it – its **weight** (downwards) and **air resistance** or **drag** (upwards).
- When an object is dropped, the only force acting on the object is its weight and hence it **accelerates** at 9.8 m/s^2.
- As the speed of the falling object increases, the resistance due to air also increases. This decreases the acceleration of the object.
- After some time, the drag is equal to the weight of the object. The **resultant force** is zero. The acceleration is also zero and the object has a constant velocity known as **terminal velocity**.
- For an object falling through a fluid (gas or liquid), the drag force is directly proportional to the square of the speed. That is: drag $\propto$ speed^2.
- $$\text{acceleration} = \frac{\text{weight} - \text{drag}}{\text{mass}}$$

Now try this

a Are the following statements **true** or **false**?

When an object is dropped, the only force acting on it is its weight. _____

An object falling through air has an upward force called drag. _____

Drag decreases as the velocity of the object increases. _____

A stone of weight 1.2 N reaches terminal velocity. The drag is 1.2 N. _____

TOP TIP A falling object will accelerate as long as the weight and drag are unbalanced.

Homework

1 Make a list of objects that at some time have a terminal velocity.

2 Use your knowledge of physics to describe the motion of a skydiver jumping from a tall building.

3 Use the Internet to find the terminal velocity of a skydiver.

CAR SAFETY

- The **thinking distance** is the distance travelled before the brakes are applied. It increases if the driver is tired, has taken drugs or alcohol, or the speed of the car is greater.
- thinking distance = speed of car × reaction time of driver
- The **braking distance** is the distance travelled by a car as the brakes are applied and the car stops. It increases when the road is wet or icy, the brakes are worn, the tyres are bald and the speed of the car is greater.
- **stopping distance** = thinking distance + braking distance
- Car safety features are **seatbelts**, **airbags** and **crumple zones**.
- Seatbelts prevent collisions with the dashboard and windscreen. They increase the person's time for stopping and therefore decrease the person's deceleration and the force acting on them.
- Airbags inflate rapidly during a collision, preventing collisions with the dashboard and steering wheel. They also increase the person's time for stopping and therefore decrease the person's deceleration and the force acting on them.
- The crumple zones protect the passengers by absorbing some of the energy of the collision.

Now try this

b Fill in the missing words.

The reaction time of a driver is 0.6 seconds. The thinking distance when the car is travelling at a speed of 10 m/s is _____m. Wet conditions will increase the _____ distance and tiredness increases the _____ distance. Driving in wet conditions and being tired will increase the overall _____ distance of the car.

Homework

4 Describe the factors that affect the braking distance of a car.

5 Describe how seatbelts and airbags prevent injuries during a car crash.

6 Use the Internet to research and write about the safety features in a particular car.

TAKE A CHANCE

- For sports there are several ways of assessing the risks: total number of deaths; number of people treated in hospitals; number of deaths as a percentage of participants in a sport.
- The risk of being killed by falling coconuts is greater than the risk of being killed by a shark.
- People feel safer in cars because they are familiar with cars and feel in control.
- People feel less safe in trains and planes because someone else is in control.

Now try this

c Circle the correct answers.

i We can assess risks in a particular event by determining the number of:

births   deaths

ii People feel safer travelling in cars because they are:

in control   travelling fast

Homework

7 Describe how skydivers or deep-sea divers minimise the risks to their lives.

8 More people are killed in car accidents than in aeroplanes. Why do people feel safer in cars?

Work and energy

WORK AND POWER

- **Work done** is measured in **joules** (J).
- Work is the amount of energy transferred when a force moves an object through a certain distance.
- work done = force × distance moved in the direction of the force

  $W = F \times s$
- 1 joule is the work done on an object when a force of 1 N moves the object a distance of 1 m in the direction of the force.
- **Power** is measured in **watts** (W).
- Power is equal to the rate of work done.
- power = $\dfrac{\text{work done}}{\text{time}}$ or power = $\dfrac{\text{energy transfer}}{\text{time}}$
- 1 W is equal to 1 J per second.

Now try this

a Match the beginnings and endings to make complete sentences.

Beginning	Ending
No work is done on an object if it	joules.
Power is measured in	work done.
Work is measured in	remains stationary.
In the equation: $W = F \times s$, W is	watts.

Homework

1 Make a list of all the quantities in this section and their associated units.

2 In your own words, define work done and power.

ENERGY

- The **gravitational potential energy** of an object increases as it gains vertical height.
- potential energy = mass × gravitational × change in
       transfer   field strength   height

  $PE = mgh$ (g = 10 N/kg on the Earth's surface)
- A moving object has **kinetic energy**.
- kinetic energy = $\frac{1}{2}$ × mass × velocity^2
  $KE = \frac{1}{2}mv^2$
- Potential energy and kinetic energy are both measured in **joules**.
- **electrical power** = voltage × current
  $P = VI$
- For a motor:
  electrical energy transfer = voltage × current × time
  $E = VIt$

Now try this

b Is each statement **true** or **false**?

The potential energy of a ball changes as it rolls on a horizontal surface. _____

A car travelling at 10 m/s has less kinetic energy than when travelling at 20 m/s. _____

Doubling the speed of a car increases its kinetic energy by a factor of two. _____

The power of a motor connecting to a 12 V supply and carrying a current of 2 A is 6 W. _____

Homework

3 Make a list of all the quantities in this section and their associated units.

4 Use the Internet to find the typical mass and speed of objects. Present your findings in a table. For each object, calculate the kinetic energy in joules.

ROLLER COASTERS

- Energy is always **conserved**, that is:
  total initial energy = total final energy

- For a roller coaster:

  As it gains height: The KE decreases and the PE increases.
  loss in KE = gain in PE

  At the top of the ride: The PE is at a maximum.

  As it falls: The PE decreases and the KE increases.
  loss in PE = gain in KE

  When it comes to a stop: The KE is transferred into heat due to the brakes.

- An object moving in a circle with constant speed has a changing velocity because of the change in the direction of travel.
- An object moving in a circle has a **resultant force** and **acceleration**.
- The direction of the resultant force is always directed towards the centre of the circle. This force is known as the **centripetal force**.

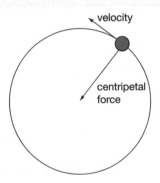

- The acceleration is always directed towards the centre of the circle. This acceleration is known as **centripetal acceleration**.
- The velocity is always at right angles to the centripetal acceleration.
- Examples of objects moving in a circle are: a roller coaster, a stone attached to the end of a string, planets orbiting the Sun, satellites orbiting the Earth, etc.

Now try this

c Tick all the correct statements below.

A satellite moving around the Earth has a constant velocity. ☐

A stone whirled in a horizontal circle has zero resultant force. ☐

A car going round a corner has acceleration. ☐

The resultant force on an object moving in a circle is called centripetal force. ☐

The centripetal force and acceleration are at right angles to each other. ☐

Homework

5 Describe the changes in energy when a monkey falls from a tall tree.

6 Design a simple roller coaster. Describe the changes in energy for a carriage on this roller coaster.

7 Using a labelled diagram show the velocity, acceleration and net force for a planet orbiting the Sun.

Einstein

NEW IDEAS

- **Einstein** was awarded the Nobel Prize for physics in 1921. He developed the **Special Theory of Relativity**.
- Einstein came up with creative ideas that challenged long-established ideas of scientific theories of famous people like **Newton**.
- Einstein carried out 'thought experiments' and made some very specific predictions.
- Most of these predictions seemed to go against common sense as well as established scientific theories.
- These predictions were tested by experiments.
- Experiments have confirmed Einstein's theories to be true.
- Einstein's ideas improved our understanding of the world around us.
- His Special Theory of Relativity was based upon the fundamental assumption that the speed of light was the same and did not depend on the relativity velocity of people or objects.
- Einstein's **General Theory of Relativity** was based on the idea that it was impossible to tell the difference between acceleration caused by gravity and acceleration caused by a force as given by $F = ma$.

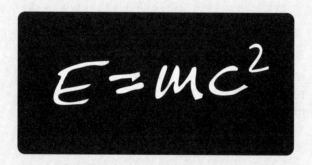

$$E = mc^2$$

Now try this

a Find the following key words in this word search.

theory	Einstein	light
thought	test	special

N	S	P	E	C	I	A	L
E	V	E	L	I	G	H	T
E	A	P	F	A	S	G	H
N	I	E	T	S	N	I	E
T	L	A	U	E	T	W	O
U	T	P	C	T	S	A	R
K	E	U	O	K	J	T	Y
A	T	H	G	U	O	H	T

 TOP TIP Random ideas do not automatically produce new ideas.

Homework

1 Describe how the work of Einstein differed from earlier scientists.
2 You have devised a certain theory. Outline what you have to do to have your ideas accepted.
3 List some of the predictions made by people on a daily basis.

EINSTEIN'S THEORIES

- Here are some of the predictions made by Einstein's **Theory of Special Relativity**:
  - The **speed of light** is the same everywhere.
  - Faster moving objects have a greater mass than when at rest (for example, an electron travelling close to the speed of light can have a mass 20 times greater than its 'rest' mass).
  - Moving clocks run slower than stationary clocks. This is called **time dilation**.
  - The evidence for time dilation comes from short-lived cosmic particles known as muons that are produced in the upper regions of our atmosphere. They last longer when travelling at high speeds.
  - Further evidence for time dilation comes from accurate **atomic clocks**. An atomic clock runs slower when carried on fast aeroplanes.
  - Moving objects shrink in length when travelling at speed close to that of light. This is known as **length contraction**.
  - Mass and energy are related by the equation: $E = mc^2$ (where E is the change in energy, m is the change in mass and c is the speed of light in a vacuum (3.0×10^8 m/s)).

- Here are some of the predictions made by Einstein's **General Theory of Relativity**:
  - The gravity of a star or galaxy curves the space around it.
  - The curved space bends starlight passing close to a dense star or galaxy.
  - The curved space explains the strange orbit of Mercury, which is closest to the Sun. (The orbit of Mercury 'twists' over time.)

Now *try this*

b Tick the statements that are predictions made by Einstein.

Moving clocks run slower. ☐

Moving electrons become more massive. ☐

Cosmic particles known as muons provide the evidence for time dilation. ☐

For an accelerating object, the force is given by $F = ma$ ☐

Energy and mass are related by $E = mc^2$. ☐

The speed of light is the same everywhere. ☐

Homework

4 Draw a mind map for Einstein's Special Theory of Relativity.

5 Outline the evidence that supports the theory that 'moving clocks slow down'.

6 Use a search engine to find some images of starlight bent by gravity. Make a list of websites for the class.

Radioactivity

ATOMS, NUCLEI AND ISOTOPES

- An **atom** has a positive **nucleus** that is surrounded by orbiting negative **electrons**.
- The nucleus is very massive and is 10 000 times smaller than the atom.
- The nucleus contains **protons** and **neutrons**. Protons have a positive charge and neutrons have no charge.
- The protons and neutrons are also known as the **nucleons**.
- The nucleus of an atom may be represented as $^A_Z X$, where X is the **chemical symbol**, A is the nucleon or **mass number** and Z is the proton or **atomic number**.
- The **isotopes** of an **element** are nuclei that have the same number of protons but a different number of neutrons.

carbon-14 $^{14}_6 C$

Now try this

a Circle the correct answers.

  i A nucleus of an atom is:

    negative   neutral   positive

  ii An example of a nucleon is a:

    nucleus   electron   proton

  iii All isotopes must belong to the same:

    element   atom   material

 TOP TIP A neutral atom has the same number of protons and electrons.

Homework

1. Write a short paragraph on the atom.
2. Use the Internet to find some isotopes of **two** elements. List the isotopes for the class.

BACKGROUND RADIATION

- **Background radiation** is either 'natural' or 'artificial'.
- Most background radiation comes from natural sources: the Sun, outer space, rocks and food.
- Artificial sources of background radiation are: nuclear power stations, fallout from nuclear explosions or accidents, waste from hospitals and X-rays.
- Some rocks, like granite, emit radioactive gas called **radon**.
- Radon trapped in houses can cause cancer.
- Some regions (for example, Cornwall and Scotland) are more prone to the dangers of radon gas than others.

 TOP TIP You can reduce the dangers of radon gas by ventilating your house.

Now try this

b Fill in the missing words.

Exposure to radiation can lead to _____. Rocks such as _____ emit the radioactive gas called _____ naturally. Even the food we eat is slightly radioactive. This is not dangerous because the levels of radiation are quite _____. Most background radiation is not caused by humans, but is created _____ by our surroundings.

Homework

3. List all the contributors to background radiation.
4. Describe the dangers of radon gas.

RADIOACTIVE DECAY RULES

- Unstable nuclei emit either **alpha (α)** or **beta (β) particles** and/or **gamma (γ) rays**.
- All three radiations cause **ionisation** – they can strip off electrons when colliding with atoms.
- An alpha particle is a **helium nucleus** ($^4_2$He). It is positive and can be stopped by a thin piece of paper.
- A beta particle is an **electron**. It is negative and can be stopped by a few millimetres of aluminium.
- Gamma rays are short-wavelength **electromagnetic waves**. These have no charge and can be stopped by a few centimetres of lead.
- The **activity** of a source is the number of nuclei decaying per second. Activity is measured in **becquerels (Bq)**.
- The **half-life** of an **isotope** is the average time taken for half the active nuclei to **decay** or disintegrate.
- Radioactive decay is a random process because it is impossible to predict which nucleus will decay at a particular time.

Now try this

c Tick the correct statements.

Alpha radiation consists of particles. ☐

Beta particles can be stopped by lead. ☐

Gamma rays can be stopped by a thin piece of paper. ☐

Three-quarters of the nuclei decay after two half-lives. ☐

Homework

5   Describe the nature of alpha particles, beta particles and gamma rays.

6   A particular isotope has a half-life of 2 hours. With the help of a graph, explain how the activity of a source would change with time.

7   Write a paragraph to explain to a GCSE student what is meant by 'radioactivity is a random event'.

USES OF RADIATION

- Irradiating fresh fruit (strawberries) with **gamma rays** prolongs their shelf life because gamma rays kill bacteria and microbes.
- A smoke detector uses an alpha-emitting source with **americium-241**.
- In **radioactive dating**, the age of rocks can be found by determining the number of **uranium nuclei** transforming into lead nuclei.
- **Carbon-dating** can be used to date bone, cloth, wood and paper. The ratio of carbon-14 to carbon-12 is used to find the age of a relic.

Now try this

d Match each comment on the left-hand side with the correct term on the right.

i   Gamma rays kill off   nuclei

ii Smoke detectors use

iii The isotope of americium   radioactive
    has 241

iv The carbon-14 isotope is   americium-241

v   Age of rocks can be   bacteria
    found from the decay of
    uranium   nucleons

Homework

8   Write a paragraph on the uses of radiation.

9   Use the information provided on page 247 of Collins GCSE Additional Science to explain carbon dating.

10 Write a short paragraph to explain why irradiating food with gamma rays keeps them fresh for longer.

Using radiation

X-RAYS AND GAMMA RAYS

- **X-rays** and **gamma (γ) rays** are both electromagnetic waves. In a vacuum, they travel at the same speed, c, equal to 300 000 000 m/s.
- Gamma rays have a shorter wavelength than X-rays.
- The speed, c, of an electromagnetic wave is related to its **frequency**, f, and **wavelength**, λ, by the equation: $c = f\lambda$.
- X-rays and gamma rays are both **ionising radiations**. Gamma rays are more ionising than X-rays.
- Large dosages of X-rays or gamma rays can cause **cancer**.
- Denser bone stops X-rays, but they pass easily through the softer tissue.
- Gamma rays are emitted by radioactive nuclei (for example, cobalt-60).
- X-rays are produced when high-speed electrons hit a target metal (for example, tungsten).

Now try this

a Tick the correct column for each statement.

Statement	X-rays	Gamma rays
i Can cause cancer	☐	☐
ii Produced by radioactive nuclei	☐	☐
iii Produced by electrons striking a metal at high speed	☐	☐
iv Travel through a vacuum	☐	☐
v Stopped by dense bones	☐	☐

Homework

1 Describe the differences and similarities between X-rays and gamma rays.
2 Use a search engine to find X-ray images. List **three** interesting websites for the class.
3 Write a short paragraph to explain how X-rays are produced.

MEDICAL USES OF RADIOACTIVITY

- **Gamma rays** are used to **sterilise** bandages, syringes and other hospital instruments.
- Gamma radiation kills dangerous bacteria like staphylococcus.
- In **radiotherapy**, several gamma ray sources are directed towards cancerous tissues to destroy the cancer cells.
- The function of some vital organs can be diagnosed using a **radioactive tracer**.
- A radioactive tracer is a radioactive substance emitting gamma radiation that is injected into a patient. The gamma rays are detected using a **gamma camera**.
- The most common tracer is **technetium-99**, used to diagnose the functions of the lung, heart and brain.

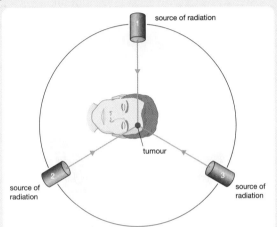

source of radiation

tumour

source of radiation

source of radiation

Now try this

b Find six key words in this word search.

A	Q	L	K	I	T	S	F
G	D	G	R	O	U	S	G
O	B	A	C	N	C	T	Q
M	D	M	E	I	V	E	T
G	X	M	G	S	C	R	Y
T	R	A	C	E	R	I	O
S	A	J	L	H	X	L	P
M	Y	L	J	D	D	E	E

TOP TIP Hospital syringes come in sealed sterile plastic bags. Exposure to gamma rays sterilises both the bag and the syringe.

Homework

4 Draw a mind map for tracers.
5 Write a short paragraph on medical uses of gamma rays.
6 Write a short paragraph on radioactive tracers.

DANGERS OF RADIATION

- All ionising radiations (**alpha particles**, **beta particles**, **gamma rays** and **X-rays**) are dangerous.
- Ionising radiations can destroy healthy cells. They can also damage the DNA that can lead to mutation of cells and cancer.
- The Earth's atmosphere and its magnetic field (magnetosphere) protect us from the harmful energetic charged particles from outer space and the Sun.

Now try this

c Fill in the missing words.

All ionising radiations destroy healthy _____ and can potentially cause _____. The Earth's magnetic field deflects the charged particles from outer _____ and the _____. The Earth's atmosphere also protects us by stopping the charged _____ reaching its surface.

Homework

7 Explain to a GCSE student why both X-rays and gamma rays are dangerous.
8 Predict the nature of the Earth's surface if it did not have a magnetic field.
9 Write a short paragraph to explain what is meant by ionisation.

Electrostatics

CHARGING INSULATORS

- Rubbing can charge **insulators** like wool, plastic and rubber.
- When two insulators rub together, the friction between them causes some of the outer **electrons** of the atoms to be stripped off from one of the insulators.
- An insulator acquires a **charge** by transfer of electrons.
- An insulator becomes negative when it gains electrons and positive when it loses electrons.
- Like charges **repel**; unlike charges **attract**.
- A polythene rod rubbed with a duster acquires a negative charge and the duster acquires an equal but opposite positive charge.

Now try this

a Tick the correct statements.

An insulator becomes positive because it gains protons. ☐

All insulators can be charged by friction. ☐

Two positively charged balloons attract each other. ☐

Two electrons will repel each other. ☐

Homework

1 Draw a mind map for 'charging insulators'.

2 Charge a plastic ruler by rubbing it with a cloth. Explain how the ruler acquires a charge.

3 Make a list of items in the home that suffer from electrostatic dust.

ELECTRIC SPARKS

- As the charge on an isolated insulator increases, the **voltage** (or **potential difference**) on the insulator relative to the ground also increases.
- When the voltage on an isolated insulator becomes too large then electrons and other charges can be transferred through the air as an **electric spark**.
- Electric sparks can generate high temperatures and therefore are dangerous near petrol fumes.
- **Earthing** an object can reduce electric sparks.
- A lightning conductor on top of a tall building protects it from a lightning strike.

 TOP TIP You can reduce electric sparks by securing a metal strap between the object and the ground. This helps to transfer charges to the Earth.

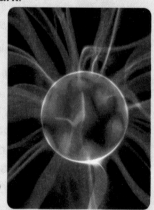

Now try this

b Complete the sentences by filling in the missing words.

An electric _____ is created when charges are transferred between an object and the ground through the _____. Electric sparks are very _____ because they increase the temperature of the air. A tall building is protected from dangerous lightning strikes by having a metal _____ placed at its highest point. If there is a lightning strike, then charges are safely transferred to the _____.

Homework

4 Make a list of places where electric sparks can be potentially very dangerous.

5 Electric sparks can be dangerous but they do have some uses. Suggest some of the benefits of electric sparks.

THE USES OF ELECTROSTATICS

- In coal-burning power stations, smoke particles (soot) in the chimney are removed by first charging them and then attracting them to oppositely charged metal plates.
- Placing paper on charged wires and then sprinkling black powder can detect fingerprints on the paper. The powder sticks to the fingerprint and not the clean paper.
- **Electrostatic** attraction is used in laser printing (photocopier).
- This is how laser printing works:
  - The image is projected onto a drum made of selenium.
  - Selenium is sensitive to light. Only the dark sections of the drum become positively charged.
  - Fine black powder (toner) is charged negative and this sticks to the positive sections of the drum.
  - The black powdered image is transferred to paper by contact.
  - The image is 'fixed' on the paper by heating it.

Now try this

c Find the following key words linked to laser printing in this word search.

light   heating   image
powder   laser   drum

W	E	R	K	A	V	A	L
T	H	G	I	L	Z	D	L
X	E	L	T	V	D	P	A
N	A	D	R	U	M	O	S
E	T	H	W	G	F	W	E
E	I	F	X	P	U	D	R
T	N	I	M	A	G	E	B
U	G	C	S	A	S	R	J

 TOP TIP Remember that opposite charges attract.

Homework

6 Using a diagram, explain how smoke is removed from chimneys.

7 Draw a block diagram to illustrate the stages of making a photocopy of a document.

8 Write a short paragraph to describe the effect laser printers (photocopiers) have had in schools.

Power of the atom

EINSTEIN'S CONTRIBUTION

- All great scientists like Newton and Einstein made predictions. Only experiments can confirm if a particular idea or model is correct.
- Einstein's famous mass-energy equation is $E = mc^2$, where E is the change in energy, m is the change in mass and c is the speed of light $(3.0 \times 10^8$ m/s) in a vacuum.
- In $E = mc^2$, E is in joules (J), m is in kilograms (kg) and c is 300 000 000 m/s.
- Whenever the mass of a 'system' decreases, then energy is released.
- The Sun releases energy because it converts mass into energy.

Now try this

a Circle the correct answer for each statement.

i Scientists make predictions which have to be tested by:

   predictions   experiments

ii Mass is measured in:

   joules   kilograms   metres per second

iii The mass of a moving electron is:

   larger   smaller than when at rest

iv The Sun's mass is:

   increasing   decreasing because it releases energy

Homework

1 Quote Einstein's famous mass-energy equation and explain what it means.

2 Find the mass of five everyday objects. Present your results in a table. Use $E = mc^2$ to calculate the maximum possible energy that you can get from each object.

FISSION

- In a **fission** reaction, a **neutron** is captured by a **uranium-235 nucleus** and it splits into two smaller (daughter) **nuclei** and either two or three neutrons.
- In a fission reaction, there is a decrease in mass and therefore, according to $E = mc^2$, energy is released.
- In a **chain reaction**, the neutrons can cause further fission reactions.
- A destructive use of fission is the **atomic bomb**. The chain reaction is out of control. As a result, enormous energy is released in a very short time.
- One of the peaceful uses of fission is the production of electricity in a nuclear power station. The chain reaction is controlled such that on average one neutron is captured from one reaction to the next.

Now try this

b Tick the correct statements.

i In a fission reaction, a nucleus splits into lots of neutrons. ☐

ii A uranium-235 nucleus can produce fission. ☐

iii Energy is released in a fission reaction. ☐

iv A fission reaction happens when a uranium nucleus captures a proton. ☐

v The chain reaction is controlled in an atomic bomb. ☐

Homework

3 Describe to a GCSE student why energy is released in a fission reaction.

4 Use a labelled diagram to explain what is meant by nuclear fission.

5 Write a short paragraph to describe the peaceful and destructive nature of fission.

NUCLEAR REACTOR

- In a **nuclear reactor**, heat is produced from the **fission** reactions.
- A nuclear reactor has:
  - **fuel rods** made from uranium-oxide
  - **control rods** (made from cadmium or boron) that absorb the neutrons in order to control how quickly the fission reactions occur
  - **moderator** (either graphite (carbon) or water) used to slow the fast neutrons
  - shielding to prevent the radiation from the nuclear core escaping.
- The waste from nuclear reactors can remain radioactive for thousands of years.
- The waste from nuclear reactors can be buried underground in remote and dry regions.

Now try this

c Match the beginnings and endings to make complete sentences.

Beginning	Ending
The fuel used in a nuclear reactor is	neutrons.
A moderator is used to slow down the	boron or cadmium.
Control rods are made from either	uranium-oxide.
In a nuclear reactor, the chain reaction is	controlled.

 TOP TIP Slower neutrons have a greater chance of being captured by uranium nuclei.

Homework

6 Write a paragraph about the disposal of nuclear waste.
7 Draw a labelled diagram of a fission reactor.
8 Describe the terms: moderator, control rods, fuel rods.

FUSION

- In a **fusion** reaction, **nuclei** of hydrogen join together to produce helium and lots of energy.
- Fusion requires high temperatures so that the fast moving nuclei can get close enough to fuse. (Remember that nuclei are positive and therefore repel each other.)
- The Sun produces its energy through fusion reactions.
- Two scientists, Pons and Fleishman, in 1989, were convinced that they had produced fusion at room temperature (cold fusion). Other scientists could not reproduce their experiments.

Now try this

d Fill in the missing words.

_____ is released in all fusion reactions. Fusion requires high _____ because the positive nuclei _____ each other. The Sun produces its energy by fusion reactions. The _____ of the Sun is converted into energy according to $E = mc^2$.

Homework

9 Explain how fusion differs from fission.
10 Suggest why we do not have fusion reactors on the Earth.

Biotechnology

USEFUL MICROBES

- **Microorganisms** can be used to make a range of foods.
- In yoghurt production, **fermenting bacteria** convert the lactose in milk into lactic acid, essentially making the milk go 'off'.
- Bacteria are also vital in the production of soy sauce – soya beans are fermented.
- **Enzymes** are **biological catalysts**. They are proteins which speed up chemical reactions.
- Enzymes are found in washing powder to break down stains.
- Enzymes such as invertase (produced by the yeast *Saccharomyces cerivisiae*) are inserted into the middle of some sweets to make a softer texture.
- Enzymes are also used in the manufacture of monosodium glutamate, vegetarian cheese and fizzy drinks.

Now try this

a Circle the conditions at which enzymes work best.

High pH

High concentration of reactant

Boiling temperature

Low pH

Specific pH for each enzyme

Freezing

37 °C

Low concentration of reactant

Homework

1 What is the name of the bacteria used to make yoghurt?
2 What do we mean by 'good' and 'bad' bacteria?
3 Write the word equation for the production of alcohol by fermentation.

PLANT MODIFICATION

- **Genetic modification** in plants is extremely useful.
- The required gene is inserted into the bacteria *Agrobacterium tumefaciens*. This then infects the plant cells causing tumours to be formed. Cells from these tumours contain the new gene and can be used to **clone** new plants.
- We can genetically modify plants to make them **resistant** to herbicides, which can then be sprayed on fields to prevent weed growth without affecting the plant itself.
- We can also genetically modify plants to contain a toxin, usually produced by *Bacillus theringiensis*, which makes the plants **insect-resistant**, thus reducing the need for pesticides.

Now try this

b Are the following statements **true (T)** or **false (F)**?

Genetic engineering is impossible in plants _____

Tumour-forming bacteria are useful as we can use them to 'infect' plant cells with useful genes _____

We use genetic modification to make plants that are resistant to herbicides _____

Genetic engineering of plants can enable farmers to use fewer pesticides _____

Homework

4 Research some genetically modified crops. Is it likely that you have eaten a genetically modified plant?
5 How might we genetically modify plants to help reduce famine in the Third World?
6 Explain why cross-pollination of genetically modified herbicide-resistant plants with wild plants may pose a problem.

REPRODUCTION

- Allowing people to decide the sex of their baby by choosing which sperm fertilises the egg could result in certain populations having more of one sex than another.

- Another worry is that if we allow parents to choose the sex of their baby, soon they will also be choosing other features such as eye colour.
- In some countries, **embryonic stem cell research** is not permitted as it is believed to be unethical to create life just for scientific research.
- Other people argue that such research has massive implications, leading to the cures of diseases such as Parkinson's.

Now try this

c Circle the qualities that could be altered with gene therapy.

Eye colour

Personality

Hair colour

Dress sense

Cystic fibrosis

Sex

Cancer

Facial scars

Homework

7 Do you believe stem cell research is ethical? Explain your answer.

8 What features do you think we should be allowed to change?

9 If we cannot change the physical features of a person before they are born through gene therapy, is it ethical to change them when they are adults through plastic surgery?

NATURAL PHARMACEUTICALS

- For thousands of years we have used natural herbal remedies as medicines:
  - Aspirin: found in the bark of the willow tree, used for pain relief
  - Taxol: from the bark of the Pacific Yew tree, used for cancer relief
  - Artemisinin: extracted from the Chinese plant *Artemsia annua*, treats and reduces the transmission of malaria
  - Quinine: from the bark of the cinchona tree, until the 1930s was the only real treatment for malaria.
- Many indigenous populations still use such treatments.
- Western medicine has a lot to learn from traditional medicine. Often, synthetic versions of a plant extract can be made and mass-produced.
- The demolition of natural areas is hugely detrimental as potential cures for diseases may be destroyed as plants become extinct.

Now try this

d Match the remedies with their sources.

Aspirin	Found in the bark of the Pacific Yew tree
Taxol	Found in the bark of the willow tree
Quinine	Extracted from the Chinese tree, Artemsia annua
Artemisinin	Extracted from the bark of the Cinchana tree

Homework

10 Why is maintaining areas such as the rainforest so important for modern medicine?

11 Research the popular Chinese herb ginseng – what is it used for?

12 How beneficial are dietary supplements? Can vitamins replace fresh fruit?

Animal behaviour 1

INSTINCTIVE OR LEARNED?

- Certain **behaviour** is **innate** – animals **inherit** these behaviour patterns from their parents. We call this **instinctive behaviour**.
- Instinctive behaviour includes **reflexes**. For example, ducking when something comes close to your head.
- Other behaviour can be learned, either from the parents or through trial and error.
- **Habituation** is an important part of the learning process. Organisms learn through repetition not to respond to a stimulus. For example, birds learn not to respond to scarecrows.
- **Conditioning** is when animals learn to respond to a certain stimulus.

Now try this

a Are the following types of behaviour **learned (L)** or **instinctive (I)**?

Riding a bicycle _____

Being wary in the dark _____

Sniffer dogs _____

Being nocturnal _____

Living in a pack _____

Reading and writing _____

Homework

1 Describe **one** example of instinctive behaviour.
2 Describe **one** example of learned behaviour.
3 Find out about Pavlov's dog. What type of learned behaviour was exhibited in this example?

SOCIAL BEHAVIOUR

- **Behaviour** requires **communication** between animals.
- Depending on the species, communication can take a number of forms – sounds, chemicals (**pheromones**), actions and facial expressions.
- Humans are different from most animals because they have developed some quite complex methods of communication.

- Humans are conscious of their actions. They have **emotions** thus making their social behaviour much more complex than most other organisms.
- Humans are also able to talk to each other, passing on important information and complex ideas.

Now try this

b Fill in the missing words.

Animals have evolved methods of _____ with each other. Communication can take a number of forms. _____ are chemicals which, when released, signal a message to other organisms. _____ expressions and _____ posture enable animals to display emotions and feelings. _____ have quite complex social behaviour as they are _____ of their actions and have _____.

Homework

4 List the ways in which animals can communicate with each other.
5 Find out how ants communicate with each other. Why might ants need to communicate?
6 Explain which factors mean that human social behaviour is very complex.

See pages 42–51 of Collins GCSE Biology

FEEDING BEHAVIOUR – HERBIVORES

- **Feeding behaviour** depends on the type of food being consumed.
- Herbivores do not catch their food. They spend most of their time grazing and must consume much more than carnivores to gain sufficient energy and nutrients.
- Herbivores have evolved teeth suitable for constant chewing and grinding.
- Vertebrate herbivores tend to graze in large herds for protection. They often have signals to warn each other of approaching predators.
- Feeding in large herds is an important evolutionary strategy. The slower (often diseased or older) members of the herd may become prey to carnivores, resulting in the survival of the fittest.
- Vertebrate herbivores tend to move over large areas, following the availability of food and water.

Now try this

c Fill in the missing words.

Herbivores must consume _____ food than _____ in order to get enough nutrients and _____ to survive. Vertebrate herbivores tend to graze in large _____. They do this for protection in _____. Herbivores have to be good at _____ away from predators.

Homework

7 Why must vertebrate herbivores be particularly good at fleeing predators?

8 Why must herbivores consume more than carnivores?

9 Explain why vertebrate herbivores have evolved to graze in large groups.

FEEDING BEHAVIOUR – CARNIVORES

- Meat is protein-rich and energy-rich. This means **carnivores** spend less time eating.
- Carnivores must be good at hunting and catching their **prey**.
- Carnivores have evolved teeth that are specialised for gripping and killing prey.
- Some carnivores have evolved to hunt in groups. Others are able to hunt on their own. Carnivores are less likely than herbivores to live in groups.
- Organisms that show **parental care** have evolved **feeding patterns** to feed their young.

Now try this

d Match the key words with the meanings.

i Heterotroph		An organism that feeds on other organisms
ii Digestion		An organism that feeds on plants
iii Carnivore		Any organism that feeds on other living organisms and cannot produce its own food
iv Solitary		Most vertebrate herbivores feed in a _____
v Herbivore		Many carnivores live quite _____ lives
vi Herd		The process by which large insoluble molecules are broken down into small, soluble ones

Homework

10 How have bats evolved to be successful predators at night?

11 Name **one** animal that hunts alone and **one** animal that hunts in a pack.

12 Find **one** diagram of a herbivore's teeth and **one** of a carnivore's teeth. Label the major differences and adaptations.

Animal behaviour 2

REPRODUCTIVE BEHAVIOUR

- **Sexual reproduction** requires the selection of a suitable **mate**.
- This often results in **courting behaviour** – a mating dance, song or fight.
- Some animals, such as swans, mate for life. Others select several different mates.
- **Survival of the fittest** means organisms will select a partner according to their **adaptations** to a certain feature. For example, stags fight over a doe. The winner mates with the female so the genes coding for the stag to be strong will be passed onto the offspring.
- Some organisms, such as insects, release special chemicals called **pheromones**.
- Male birds often have bright colourings. Female birds are more attracted to these colourings and therefore are more likely to breed.

Now try this

a Fill in the missing words.

_____ reproduction often involves the selection of the mate. This often results in _____ behaviour. This may involve a mating _____, _____ or a _____. Some organisms release _____. These courtship rituals often result in selection of the _____ male.

Homework

1 Find out about the mating rituals of snakes.
2 What is the 'rutting season'?
3 Explain how courting rituals often result in the spreading of the most advantageous genes.

REARING YOUNG

- Some animals have developed certain behaviours for **rearing** their young.
- There is an increased chance of offspring survival if they are cared for by the parent.
- If the offspring survives, the parent's genes will be successfully passed on.
- Often, if the **gestation period** is longer or few offspring are produced at once, the parent invests more time into rearing the offspring.
- The African elephant gestation period is 22 months and the offspring will stay with the mother for about 16 years.

Now try this

b Match the key words with the meanings.

i Long gestation period	The length of time the mother is pregnant
ii Offspring survival	Increased by the length of time spent by the parents rearing the young
iii Gestation period	Usually less time invested in each one
iv More offspring	Offspring usually stay with the mother for longer

Homework

4 What is the gestation period for a human?
5 What is the general relationship between gestation period and effort put into raising the offspring?
6 Why does spending time raising offspring result in an increased chance of survival for the parents' genes?

HUMAN BEHAVIOUR

- Humans are one of the 'Great Apes'. They are closely related to *bonobos* (pygmy chimpanzees).
- Humans started as a small group of **hunter-gatherers**.
- **Human behaviour** differs to other animals due to their **complex** societies and intelligence.
- Humans have exploited other animals, plants and their habitats for their own purposes.
- They have bred animals for food, working, clothing, hunting, companionship and entertainment.

Now try this

c Match the animal to the correct uses.

Food	Racehorse
Clothing	Fox hound
	Mink
Hunting	
	Circus animals
Companionship	Domestic cat
Entertainment	Cow

Homework

7 Find out about Lucy the *Australopithecus*. Why was her discovery such a major step for scientists studying the evolution of humans?

8 List the ways in which humans exploit other animals.

9 Explain why the evolution of opposable thumbs was such an important step for the evolution of intelligent life.

LOOKING AFTER OTHER ANIMALS?

- Much debate over the use of animals is now taking place.
- Animal cruelty is an offence in the UK.
- In modern day living, animals have many more rights than before.
- Some people believe that animals should have similar rights to humans.
- Many organisations have been set up in the fight against animal cruelty.
- In some countries animal cruelty is less frowned upon.

Now try this

d Tick if you agree that animals should be used:

In circus acts ☐
For food ☐
As rugs ☐
For labour ☐
As pets ☐
For clothes ☐
For riding ☐
For entertainment ☐
For racing ☐
In zoos ☐

Compare your answers with a friend.

Homework

10 What does 'anthropomorphism' mean?

11 What limits should be placed upon circuses that use animals? Explain your answer.

12 Should hunting with hounds have been banned in the UK? Find **three** arguments for and **three** arguments against hunting.

Chemical detection

ACIDS AND ALKALIS

- All acids contain **H⁺ ions** and change the colour of acid/base **indicators**.
- Concentrated acid solutions contain more H^+ ions and have lower **pH values**.
- **Alkaline** solutions contain **OH⁻ ions** and turn phenolphthalein pink.
- Alkalis release a pungent gas, NH_3, when warmed with ammonium salts.
- Acids **neutralise** alkaline solutions:
  $$H^+_{(aq)} + OH^-_{(aq)} \rightarrow H_2O_{(l)}$$
- They release H_2 in reactions with metals:
  $$Mg_{(s)} + 2HCl_{(aq)} \rightarrow MgCl_{2(aq)} + H_{2(g)}$$
- They release CO_2 in reactions with carbonates:
  $$CaCO_{3(s)} + 2HCl_{(aq)} \rightarrow CaCl_{2(aq)} + H_2O_{(l)} + CO_{2(g)}$$

Now try this

a Which pairs from the left-hand column react to make the chemicals on the right?

Magnesium	H₂O
OH⁻ ion solution	NH₃
Ammonium salt	
H⁺ ion solution	CO₂
Carbonate	H₂

 TOP TIP Strong acids like HCl have lower pHs than weak acids with the same concentration.

Homework

1 Show what happens to the ions and the pH, when HCl and NaOH react.

2 Find out why HCl has a lower pH than the same concentration of ethanoic acid.

3 Why do acid/base indicators change colour, and is it always at pH 7?

METAL IONS

- Chemical analysis is used in **forensics** and to check the purity of food, drink and drugs.
- **Qualitative analysis** finds what is present; **quantitative analysis** finds how much.
- Separate tests are needed for the positive and negative ions in an ionic compound.
- **Flame tests** can identify the **cations**: Na⁺ (yellow), K⁺ (lilac), Ca²⁺ (orange), Cu²⁺ (green).
- Sodium is a common contaminant and can mask the colours of the other ions.

Now try this

b What colour flames would you expect from powder on the shoes of workers in each of the following places?

Limestone quarry	_____
Salt mine	_____
Copper mine	_____
Fertiliser warehouse	_____
Blackpool beach	_____

 TOP TIP Solutions need to be evaporated to allow flame tests to be carried out on their solutes.

Homework

4 Explain how blue glass can help you see potassium contaminated with sodium.

5 How could you test claims that one bag of crisps was less salty than another?

6 Find out how the Bunsen burner allowed new elements to be discovered.

PRECIPITATES

- Group 1 **metal hydroxides** are soluble.
- They form **precipitates** when mixed with solutions of other metal ions. For example:
  $Cu^{2+}_{(aq)} + 2OH^-_{(aq)} \rightarrow Cu(OH)_{2(s)}$
- Group 2 hydroxides are white solids.
- Aluminium hydroxide is a white solid, which redissolves in excess alkali.
- **Transition metal hydroxides** are coloured: Cu^{2+} (blue), Fe^{2+} (green), Fe^{3+} (brown).
- When hydroxide is added to ammonium ions (NH_4^+), they release ammonia (NH_3).

Now try this

c Finish the following ionic equations.

$Cu^{2+}_{(aq)} + \underline{\hspace{1.5cm}} \rightarrow Cu(OH)_{2(s)}$

$Ca^{2+}_{(aq)} + 2OH^-_{(aq)} \rightarrow \underline{\hspace{1.5cm}}$

$Al^{3+}_{(aq)} + 3OH^-_{(aq)} \rightarrow \underline{\hspace{1.5cm}}$

$Fe^{2+}_{(aq)} + \underline{\hspace{1.2cm}} \rightarrow \underline{\hspace{1.2cm}}$

$Fe^{3+}_{(aq)} + \underline{\hspace{1.2cm}} \rightarrow \underline{\hspace{1.2cm}}$

$\underline{\hspace{1.5cm}} + 2OH^-_{(aq)} \rightarrow Zn(OH)_{2(s)}$

 TOP TIP Precipitates form when ions that can form insoluble compounds meet in solution.

Homework

7 Explain how to tell these pairs apart: Fe^{2+} and Fe^{3+}, Ca^{2+} and Li^+, Ca^{2+} and Al^{3+}.

8 How could you tell which of two solutions contained the most Cu^{2+} ions?

9 Find out what ions tap water contains in hard water areas.

NON-METAL IONS

- Cl^-, Br^- and I^- ions make **precipitates** with dilute nitric acid and silver nitrate solution. For example:
  $Ag^+_{(aq)} + I^-_{(aq)} \rightarrow AgI_{(s)}$.
  The nitric acid dissolves any OH^- ions present that would interfere with the test.
- AgI (yellow), AgBr (cream), AgCl (white).
- SO_4^{2-} ions make white precipitates with dilute HCl and barium chloride solution:
  $Ba^{2+}_{(aq)} + SO_4^{2-}_{(aq)} \rightarrow BaSO_{4(s)}$.
  The dilute acid dissolves any OH^- ions present that would interfere with the test.
- CO_3^{2-} ions release CO_2 with dilute acid.
- The CO_2 turns **limewater** milky.
- SO_3^{2-} ions release SO_2 with dilute HCl.
- Sulphur dioxide turns acidified potassium dichromate(VI) solution from orange to green.

Now try this

d Which reagent(s) from the right do you need to prove that the ion on the left is present?

Cl^-	Barium chloride
CO_3^{2-}	Dilute nitric acid
	Silver nitrate
SO_4^{2-}	Dilute HCl
SO_3^{2-}	Limewater
I^-	Acidified potassium dichromate(VI)

 TOP TIP Solutions that fizz with acid contain carbonate ions.

Homework

10 What solution gives a brown precipitate with OH^- and white with Ag^+?

11 Could you prove that a powder was a mixture of Na_2CO_3 and $CuCl_2$?

12 Write questions in the style of Homework question 10 for **five** other compounds (supply answers on the back).

Using equations

MOLES

- Amounts of chemicals are measured in grams, numbers of particles, or **moles** of particles.
- A mole of any chemical has a mass equal to the **relative formula mass** in grams.
- The relative formula mass is the sum of the **relative atomic masses** (M_r) of the atoms. For example: $SO_2 = 32 + 16 + 16 = 64$
- So: moles $= \dfrac{\text{grams}}{M_r}$ grams $=$ moles $\times M_r$
- The number in front of each formula shows the number of moles that react together. For example: $2H_2 + O_2 \rightarrow 2H_2O$

  2 moles   1 mole   2 moles
- Moles can be used to calculate the product formed. For example: CO_2 from 200 g of limestone – $CaCO_3 \rightarrow CaO + CO_2$

  1 mole $CaCO_3$ makes 1 mole CO_2

  200 g $CaCO_3$ is $\frac{200}{100} = 2$ moles

  2 moles $CaCO_3$ makes 2 moles CO_2

  2 moles $CO_2 = 2 \times 44$ g $CO_2 = 88$ g CO_2.

Now try this

a Convert the grams to moles and the moles to grams.

80 g Ca _____

4 g H_2 _____

36 g H_2O _____

68 g NH_3 _____

11 g CO_2 _____

0.5 moles Ca _____

10 moles H_2 _____

0.5 moles H_2O _____

2 moles NH_3 _____

0.1 moles CO_2 _____

Homework

1 Calculate the grams of CO_2 made by heating 800 g of $CaCO_3$ and burning 800 g of CH_4.

2 Calculate the masses of **ten** different oxides that contain 32 g of oxygen.

3 Calculate the mass of CO_2 that plants need to make a 1000 g bag of glucose.

GASES

- **Gas volumes** are measured in cm^3 or dm^3.
- A mole of gas occupies 24 000 cm^3 or 24 dm^3.
- moles $= \dfrac{dm^3}{24}$ $dm^3 =$ moles $\times 24$
- grams $= \dfrac{M_r \times dm^3}{24}$ $dm^3 = \dfrac{24 \times \text{grams}}{M_r}$
- Reacting volumes are easily calculated:

  $N_2 + 3H_2 \rightarrow 2NH_3$

  1 mole   3 moles   2 moles

  1 mole N_2 makes 2 moles NH_3

  24 dm^3 N_2 makes 48 dm^3 NH_3

  1 dm^3 N_2 makes 2 dm^3 NH_3

Now try this

b Convert the volumes to grams and the grams to volumes.

48 dm^3 He _____   0.5 g He _____

12 000 cm^3 H_2 _____   6 g H_2 _____

6 dm^3 O_2 _____   8 g O_2 _____

24 dm^3 NH_3 _____   68 g NH_3 _____

48 dm^3 CO_2 _____   11 g CO_2 _____

48 dm^3 CO _____   56 g CO _____

12 dm^3 CH_4 _____   64 g CH_4 _____

72 dm^3 Cl_2 _____   142 g Cl_2 _____

Homework

4 Find the volume of hot gas made when moles of CH_4, H_2, C_2H_4 and C_2H_5OH burn.

5 What mass of air fills a 24 000 dm^3 room? (Assume 80 per cent N_2 and 20% O_2.)

6 How much glucose could you burn, or use for respiration, with this much air?

SOLUTIONS

- **Solubility** is measured in grams dm^{-3} or $mol\ dm^{-3}$.
- The grams of **solute** can be found by evaporating the water from a known mass of solution.
- $1\ dm^3$ of water = $1000\ cm^3$ = 1000 g.
- 1 M solutions contain 1 mole of solute per dm^3.
- $mol\ dm^{-3}$ (**molarity**) $= \dfrac{moles}{dm^3} = \dfrac{grams}{M_r \times dm^3}$
- $grams\ dm^{-3} = mol\ dm^{-3} \times M_r$

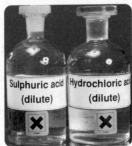

Sulphuric acid (dilute)   Hydrochloric acid (dilute)

Now try this

c Convert the grams dm^{-3} to mol dm^{-3} and vice versa.

80 g dm^{-3} NaOH	_____
63 g dm^{-3} HNO_3	_____
146 g dm^{-3} HCl	_____
7 g dm^{-3} NH_4OH	_____
49 g dm^{-3} H_2SO_4	_____
0.5 mol dm^{-3} NaOH	_____
2 mol dm^{-3} HNO_3	_____
0.2 mol dm^{-3} HCl	_____
2 mol dm^{-3} NH_4OH	_____
2 mol dm^{-3} H_2SO_4	_____

Homework

7 What mass of magnesium will react with $1\ dm^3$ of 1 M HCl or 1 M H_2SO_4?

8 Write equations for HCl and H_2SO_4 reacting with NaOH, and then $Ca(OH)_2$.

9 How could you find out whether an unlabelled acid was concentrated or dilute?

ANALYSING SOLUTIONS

- Pure water is needed for drinking and cooking.
- Water collection and purification takes time so it is important not to waste it.
- Rain water can be used for toilet flushing, etc.
- Solutions with known **molarities** are **titrated** to find the concentrations of other solutions.
- A **pipette** is used to transfer a fixed volume of acid to a flask and **indicator** is added.
- A standard solution of **base** is added from a **burette** until the indicator changes colour.
- The volume of standard solution shows the number of moles used: moles used = $mol\ dm^{-3} \times dm^3 (cm^3/1000)$.
- The equation shows what it reacts with. For example:

$Na_2CO_3 + 2CH_3COOH \rightarrow$ products
1 mole   2 moles

The unknown concentration can be found as follows:

$mol\ dm^{-3} = \dfrac{moles\ reacted}{dm^3 (cm^3/1000)}$

Now try this

d What concentration is HCl if $20\ cm^3$ is neutralised by:

$20\ cm^3$ 0.5 M NaOH	_____
$10\ cm^3$ 2 M NaOH	_____
$40\ cm^3$ 0.5 M NaOH	_____
$20\ cm^3$ 1 M Na_2CO_3	_____
$20\ cm^3$ 1 M NH_4OH	_____
$40\ cm^3$ 0.2 M Na_2CO_3	_____
$20\ cm^3$ 0.5 M $Ca(OH)_2$	_____
$10\ cm^3$ 0.010 M KOH	_____

TOP TIP Solutions with accurately known concentrations can be made from pure dry powders like Na_2CO_3.

Homework

10 Write step-by-step instructions for carrying out an acid/base titration.

11 List the sources of error in a titration and say how they can be minimised.

12 Find an image of an automatic titration machine and say what sensor it uses.

Chemical products 1

TRANSITION METALS

- **Transition metals** like iron and copper:
  - have high densities, for example osmium in pen nibs
  - have high melting points, for example tungsten in lamp filaments
  - conduct heat and electricity well, for example silver or copper.
- The metals and their compounds are:
  - **catalysts** to speed up reactions, for example Pt in catalytic converters
  - insoluble pigments in paints and glazes, for example Cu in blue colours
  - soluble dyes.

Now try this

a Link each transition metal property with its use(s).

Water pipes	Coloured compounds
Pigments	Good conductor
Dyes	Unreactive
Wiring	
Heat exchanger	Speeds up reactions
Lamp filaments	High melting point
Needles and nibs	Hard
Catalytic converter	
Electrical contacts	Non-corroding

Homework

1 Find out what effect mercury poisoning has on the body.

2 The NPI ranks cadmium the sixth most serious pollutant. Find out why.

3 Find out what phytoremediation is used for.

ACIDS, ALCOHOLS AND ESTERS

- **Alcohols** mix with water in drinks and are used as fuels and solvents, for example CH_3CH_2OH ethanol.
- They are neutral liquids, which react with sodium and **oxidise** to **organic acids**.
- Organic acids contain the –COOH group, for example CH_3COOH ethanoic acid in vinegar.
- Some flavour food or drinks, such as citric acid.
- Some are used in exfoliators to remove wrinkles and blemishes, such as glycolic acid.
- They are weak acids (partially ionised), but react with metals, hydroxides and carbonates.
- Organic acids and alcohols react (with a strong acid as catalyst) to make **esters**, for example $CH_3COOCH_2CH_3$.
- Esters have fruity smells and are used in toiletries, cosmetics and flavourings.

Now try this

b Decide if each description matches an **organic acid (O)**, **alcohol (A)** or **ester (E)**.

Contains –OH	_____
Reacts with carbonate	_____
Fruity smell	_____
Low pH	_____
Reacts with Na	_____

 TOP TIP Alcohols are like water molecules with one hydrogen replaced by a hydrocarbon chain.

Homework

4 List the names of some organic acids and esters and what they are found in.

5 Find the formula of butyric acid and the names of some foods it is found in.

6 Find out how esters like ethyl ethanoate are made in a laboratory.

OXIDATION AND REDUCTION

- **Oxidation** is loss of electrons and **reduction** is gain of electrons (OILRIG). For example:
  $Zn \rightarrow Zn^{2+} + 2e^-$ (oxidation); $Cu^{2+} + 2e^- \rightarrow Cu$ (reduction)
- **Cells** contain two metals, or a metal and another material, one connected to the positive **electrode** and one to the negative electrode.
- An **electrolyte** connects the two.
- An oxidation reaction sends electrons out at the negative electrode when the cell is in a **circuit**.
- A reduction reaction takes electrons in at the positive electrode.
- **Ions** move through the electrolyte to complete the circuit.
- In **rechargeable batteries** the reaction can be reversed by an electric current.
- In **fuel cells** the reactant is added continually while the cell is running.

Now try this

c Label the reactions **oxidation (O)** or **reduction (R)**.

Simple cell reactions

$Zn \rightarrow Zn^{2+} + 2e^-$ _____

$Cu^{2+} + 2e^- \rightarrow Cu$ _____

Fuel cell reactions

$O_2 + 4H^+ + 4e^- \rightarrow 2H_2O$ _____

$2H_2 \rightarrow 4H^+ + 4e^-$ _____

Alkaline cell reactions

$2NH_4^+ + 2MnO_2 + 2e^- \rightarrow$
$Mn_2O_3 + 2NH_3 + H_2O$ _____

$Zn + 2OH^- \rightarrow ZnO + H_2O + 2e^-$ _____

Homework

7 Produce a poster to show how one sort of cell works.

8 Explain how rechargeable car batteries and fuel cells work.

9 Find out why Duracell lets customers return their dead batteries.

ELECTROLYSIS

- **Electrolysis** decomposes **electrolytes**, which contain molten or dissolved **ions**, if inert carbon electrodes are used. (See the 'Electrolysis' panel on page 69 to review this.)
- A **battery** or **DC supply** pushes electrons to the negative electrode where they cause **reduction**:
  $Cu^{2+}_{(aq)} + 2e^- \rightarrow Cu_{(s)}$.
- **Oxidation** releases electrons at the positive electrode:
  $2Cl^-_{(aq)} \rightarrow Cl_{2(g)} + 2e^-$
  These travel through wires to the negative electrode.
- **Ions** in the electrolyte move to the electrodes to complete the circuit.
- When copper is purified the impure copper forms the positive electrode and dissolves: $Cu_{(s)} \rightarrow Cu^{2+}_{(aq)} + 2e^-$.
- The Cu^{2+} ions are attracted to the negative electrode but the uncharged impurities fall to the bottom.
- Reduction makes copper deposit at the negative:
  $Cu^{2+}_{(aq)} + 2e^- \rightarrow Cu_{(s)}$.

Now try this

d Decide if the following happen at the **positive (P)** or **negative (N)** electrodes.

Metals deposited _____

Wire collects electrons _____

Cu^{2+} ions attracted _____

Non-metals discharged _____

Reduction _____

Negative ions attracted _____

Electrolyte collects electrons _____

Oxidation _____

 The current in wires is carried by electrons. In electrolytes it is carried by ions.

Homework

10 Prepare a presentation to explain copper purification.

11 Titanium is the second most common metal – how is it extracted?

12 Find out how electroplating works and what it is used for.

Chemical products 2

ALKALI METALS

- **Alkali metals** are not like transition metals. They are soft enough to cut with a knife and have low melting and boiling points.
- Li, Na and K react with water. For example:
$$2Na_{(s)} + 2H_2O_{(l)} \rightarrow 2NaOH_{(aq)} + H_{2(g)}.$$
- **Reactivity** increases down the group, because atoms lose electrons more easily when they are further from the nucleus.

 TOP TIP The alkali metals are stored in oil to prevent them reacting with air and water.

Now try this

a Is each of the following a **transition metal**, an **alkali metal**, or **both**?

Has coloured compounds _____

Has a high melting point _____

Has a low density _____

Good conductor _____

Unreactive _____

Hard _____

Makes alkalis _____

Highly reactive _____

A good catalyst _____

Homework

1 Predict caesium's properties and then look them up and assess your accuracy.
2 Find out what superoxides are and why old pieces of potassium are a hazard.
3 Find out how sodium behaves in liquid ammonia. What is it used for?

ALKALI METAL PRODUCTS

- **Sodium carbonate** (Na_2CO_3) is melted with sand and limestone to make glass.
- $Na_2CO_3.10H_2O$ is washing soda. It is alkaline so it dissolves grease.
- **Sodium hydroxide** converts fats to soaps:
  fat → sodium salts of fatty acids + glycerol.
  These sodium salts are soap.
- NaOH breaks down the lignin between wood fibres, so the separated fibres can be converted to paper.

Now try this

b Match each item on the left to a description.

i Sodium hydroxide   Alkali

ii Sodium salt of fatty acid   Soap

       Made from fibres

iii Paper   Links wood fibres

iv Lignin   Washing soda

       Corrosive

v Sodium carbonate   Dissolves grease

 TOP TIP Concentrated sodium hydroxide is very corrosive and digests organic materials.

Homework

4 Explain the safety precautions needed when sodium hydroxide is handled.
5 Find out how soaps were made before sodium hydroxide was available.
6 Could the NaOH used to break down lignin be replaced by enzymes?

SULPHURIC ACID

- Millions of tonnes of **sulphuric acid** are made every year in the **Contact process**.
- Molten sulphur is sprayed into a furnace: $S + O_2 \rightarrow SO_2$.
- SO_2 is oxidised to SO_3 at 450 °C and 2 atm. in the presence of a vanadium oxide catalyst: $2SO_2 + O_2 \rightleftharpoons 2SO_3$.
- The conditions are a compromise.
- **Equilibrium yields** are higher at higher pressures and lower temperatures but these are too expensive and too slow to be economical.
- SO_3 is dissolved to form H_2SO_4.
- Sulphuric acid's main uses are for making:
  - ammonium sulphate fertilisers
  - dyes and pigments
  - soaps and detergents
  - fibres and plastics.

 TOP TIP If SO_3 is dissolved directly in water a mist of acid droplets forms, so it is dissolved in concentrated H_2SO_4, then diluted.

Now try this

c Fill in the missing words.

Sulphur is burned to make _____. This is mixed with more _____ and passed over a catalyst called _____ at _____ °C and _____ atmospheres. The _____ formed is dissolved in _____ sulphuric acid and then diluted to give _____ sulphuric acid.

Homework

7 Produce a poster to show some uses of sulphuric acid.

8 Find out how sulphuric acid is converted to fertiliser.

9 Research the reaction between sulphuric acid and sugar.

SOAPS AND DETERGENTS

- Oils and fats are **esters** of acids and glycerol.
- Boiling with NaOH breaks the ester linkage to make sodium salts of fatty acids for **soap**.
- Soap molecules have **ionic heads** (**hydrophilic**), which attract water molecules, and **hydrocarbon tails** (**hydrophobic**) which attract grease.
- Salts of synthetic acids are **detergents** and do not form scum with Ca^{2+} ions in hard water.
- Soaps and detergents are **surfactants**. They cut surface tension and hold dirt in suspension.
- Biological detergents have **enzymes** to digest proteins, fats and starch and dissolve stains.
- Non-biological detergents contain bleaches and usually work best at higher temperatures.

 TOP TIP Surfactants lower water's surface tension.

Now try this

d Match the description(s) on the right to each word.

Soap	In biological detergents
	Sodium salt of synthetic acid
Hydrophobic	Repels water molecules
Bleach	Sodium salt of fatty acid
Enzymes	Can form scum
Hydrophilic	In non-biological detergents
	Surfactant
Detergent	Attracts water molecules
Oil	Makes water hard
Ca^{2+}	Ester of glycerol

Homework

10 Prepare a presentation to show how detergents work.

11 Find out how water can be softened to stop scum forming.

12 What advantages and disadvantages do non-anionic surfactants offer?

Particles in action

GASES

- In physics, we use the **Celsius** and **Kelvins** scales of temperature.
- The temperature at **absolute zero** is −273 °C or 0 K.
- At absolute zero, the **kinetic energy** of the atoms or molecules is zero.
- A change of 1 °C is equal to a change of 1 K.
- The kinetic energy of gas atoms increases as the temperature increases because the particles move faster.
- The mean kinetic energy of gas atoms is directly proportional to the kelvins temperature of the gas.
- A gas exerts **pressure** because of continuous collisions of the gas atoms with the container walls.
- For a fixed amount of **ideal gas**:

$$\frac{\text{pressure}}{\text{temperature (in K)}} = \text{constant}$$

Or $\dfrac{P}{T} = \text{constant}$

- For a fixed amount of **ideal gas**: $\dfrac{P_1 V_1}{T_1} = \dfrac{P_2 V_2}{T_2}$

The temperatures must be kelvins.

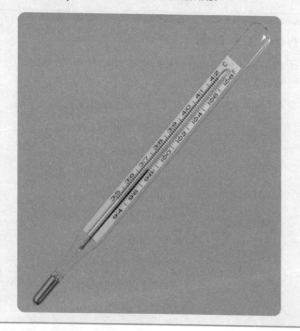

TOP TIP To convert temperature from the Celsius scale to the Kelvins scale, just add 273. Hence 20 °C is equal to (273 + 20) = 293 K.

Homework

1 Explain what is meant by 'absolute zero'. Find today's temperature in °C and show all the stages of converting it into kelvins.

2 Using a sketch graph, explain how temperature is linked to pressure.

3 Explain what is meant by 'temperature of a body'.

PARTICLES

- The **nucleus** has positively charged **protons** and uncharged **neutrons**.
- An **isotope** that has too many neutrons undergoes **β⁻ decay** (electron).
- In β⁻ decay, a neutron changes into a proton and an electron.
- An isotope that has too few neutrons undergoes **β+ decay** (positron).
- In β⁺ decay, a neutron changes into a proton and a positron.
- A positron has the same mass as an electron but an equal but opposite charge.
- A positron is the **anti-particle** of an electron.
- An isotope with more than 82 protons usually undergoes α **decay** (helium nucleus $^4_2$He).
- In α decay, the nucleon number decreases by 4 and proton number decreases by 2.
- **Quarks** and electrons are fundamental particles.
- In β⁻ decay, the nucleon number remains the same and the proton number increases by 1. This is because a down quark changes into an up quark.
- In β⁺ decay, the nucleon number remains the same and the proton number decreases by 1. This is because an up quark changes into a down quark.

Now try this

b For each particle, tick the correct column.

Particle	Fundamental	Not fundamental
electron	☐	☐
proton	☐	☐
neutron	☐	☐
up quark	☐	☐
positron	☐	☐

Homework

4 Write a short paragraph on fundamental particles.

5 Define α decay, β⁻ decay and β⁺ decay.

ELECTRONS

- **Electrons** are 'boiled off' from hot metals (filaments) by **thermionic emission**.
- kinetic energy = electronic × accelerating
   of electron   charge   voltage
   KE = eV   ($e = 1.6 \times 10^{-19}$ C)
- **Electron beams** can be deflected by electric and magnetic fields.
- Electron beams are used in TV tubes, computer monitors, oscilloscopes and X-ray tubes.
- In **particle accelerators**, high-speed particles (electrons, protons, etc) are collided with matter to learn about the fundamental nature of forces and matter.

Now try this

c Fill in the missing words.

In an X-ray tube, high-speed electrons collide with a metal target to produce _____. Doubling the accelerating _____ can double the electron's energy. Electrons can be deflected by an electric field because they carry a _____.

Homework

6 Make a list of applications where electrons are accelerated to high speeds.

7 Use the Internet to find some images of particle accelerators. Make a list of **three** useful websites for the class.

Medical physics

LOOKING INSIDE

- **Refraction** is the bending of light caused by the change in the speed of light as it travels from one medium into another.
- Light in an **optical fibre** undergoes **total internal reflection**.
- Optical fibres are used in **endoscopes**.
- An endoscope is used to send light along optical fibres and return an image so that it is possible to view inside the body.
- **work done** = force × distance moved in the direction of the force
- **power** = $\dfrac{\text{work done}}{\text{time taken}}$
- **frequency** = $\dfrac{1}{\text{period}}$ $f = \dfrac{1}{T}$
- Frequency is measured in **hertz** (Hz).
- **intensity** = $\dfrac{\text{incident power}}{\text{area}}$ $I = \dfrac{P}{A}$
- Intensity is measured in watts per square metre (W/m^2).
- The intensity of radiation (for example, light, X-rays, gamma rays) decreases with distance and the thickness of material it travels through.
- BMA stands for **basal metabolic rate**.
- Muscle cells generate **potential differences** (voltages).
- **Action potentials** can be measured with an electrocardiogram (ECG) to monitor the action of the heart.

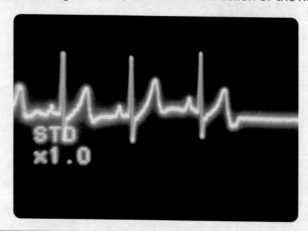

Now try this

a Are the following statements **true** or **false**?

Optical fibres use refraction of light to work. _____

Optical fibres are used in an electrocardiogram. _____

frequency × period = 1 _____

The intensity of light from a laser is measured in watts. _____

Muscles generate potential differences. _____

Homework

1 Define: work done, power, frequency, period and intensity.
2 Write a short paragraph on endoscopes and the physics behind them.
3 Use a search engine to find some ECG traces. Make a list of **two** websites for the class.

PHYSICS IN MEDICINE

- **momentum** = mass × velocity
- Momentum is conserved in all collisions and interactions between objects or particles.
- **Conservation of momentum:**
    total initial momentum = total final momentum
- Bombarding certain elements with positively charged protons makes them into **radioactive isotopes** that usually emit **positrons**.
- A positron is an anti-particle of an electron. When they meet, they immediately destroy or annihilate each other.
- When electron-positron annihilate each other, **gamma rays** are produced.
  The mass of the particles is converted into gamma rays according to Einstein's equation, $E = mc^2$.
- Both energy and momentum are conserved when a positron interacts with an electron.
- **Positron emission topography (PET)** uses the gamma rays from electron-positron annihilations to produce detailed 3-D images of the body.
- High frequency radiation (for example, X-rays and gamma rays) is dangerous because it destroys living cells.
- In **radiation therapy**, intense gamma rays or X-rays can be used in the treatment of malignant tumours.
- Radiation treatment does not always lead to a cure. It is sometimes used to reduce suffering – this is known as **palliative care**.

Now *try this*

b Tick each correct statement.

i Momentum is conserved when an electron and a positron destroy each other. ☐

ii Gamma rays destroy only cancer cells. ☐

iii A positron is another name for an electron. ☐

iv Palliative care is when suffering is reduced using radiation treatment. ☐

v The annihilation of electrons and positrons is used in PET. ☐

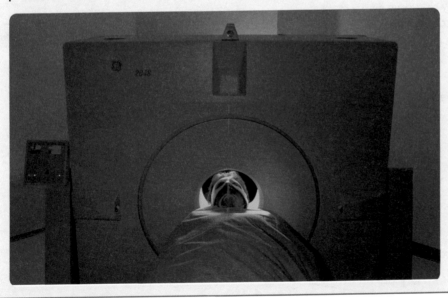

Homework

4 Explain what is meant by radiation therapy and palliative care.

5 Write a short paragraph on PET.

6 Use a search engine to find some medical 3-D PET images. Make a list of **two** websites for the class.

Exam-style questions

Biology B1a and B1b

1 A pyramid of number shows ...
- a the mass of all the animals in a food web
- b the flow of energy through a food chain
- c how many species there are at each trophic level
- d the mass of the species at each trophic level   [1]

2 Genetic modification of plants is useful for farmers because ...
- a pesticide-resistant plants can be created
- b plants can be made to grow more slowly
- c herbicide-resistant plants can be created
- d plants can be made to grow without carbon dioxide and water   [1]

3 Charles Darwin's theory states that ...
- a the least fit survive, mate and pass on their genes
- b the fittest are less likely to survive and pass on their genes
- c the fittest are most likely to survive, find a mate and pass their genes onto their offspring
- d organisms least adapted to their environment will pass on their genes   [1]

4 Asexual reproduction involves ...
- a the fusing of two gametes
- b two parents
- c fertilisation
- d only one parent   [1]

5 The blood is made up of ...
- a carbon dioxide, nutrients, oxygen and hormones
- b red blood cells, white blood cells, platelets and plasma
- c water
- d hormones   [1]

6 If blood glucose levels are too high ...
- a more glucose is released into the blood system
- b glucagon is released
- c less insulin is produced
- d more insulin is released   [1]

7 Immunity is the body's ability to resist disease; it can be acquired in three different ways. Which of the following is not a way in which immunity can be acquired?
- a Vaccination
- b Sexual intercourse
- c Mother's breast milk
- d Through being exposed to the disease   [1]

8 Which two statements are true?
- a Tuberculosis is caused by a virus
- b Tuberculosis is prevalent in people living in overcrowded conditions
- c Tuberculosis cannot be treated with antibiotics
- d Tuberculosis is airborne, transmitted through sneezes and coughing   [1]

Chemistry C1a and C1b

1 If two atoms are from the same element, they have ...
- a the same appearance
- b the same mass
- c the same proton number
- d the same neutron number   [1]

2 How many atoms are there in ammonium sulphate, $(NH_4)_2SO_4$?
- a 4
- b 10
- c 15
- d 18   [1]

3 Which of the following could be potassium chlorate?
- a KCl
- b KCL
- c $KClO_4$
- d $KCLO_4$ [1]

4 The citric acid in lemons has the formula $C_6H_8O_7$. The formula of manufactured citric acid is ...

a simpler

b more complicated

c the same

d different each time [1]

5 Citric acid neutralises carbonates. During the reaction ...

a the mixture gets hot

b carbon dioxide is released

c hydrogen is released

d citric acid gets more sour [1]

6 Acids can also neutralise oxides. A correct equation for this is:

a $H_2SO_4 + CuO \rightarrow CuSO_4$

b $HCl + CuO \rightarrow CuCl + H_2O$

c $H_2SO_4 + CuO \rightarrow CuSO_4 + H_2$

d $2HCl + CuO \rightarrow CuCl_2 + H_2O$ [1]

7 Large alkane molecules are ...

a harder to ignite

b paler in colour

c less viscous

d more volatile [1]

8 It is dangerous to restrict the oxygen supply to a gas burner because ...

a more energy is released

b carbon dioxide is formed

c carbon monoxide may form

d complete combustion takes longer [1]

9 The equation for complete combustion of CH_4 is:

a $CH_4 + O_2 \rightarrow 2H_2O + CO_2$

b $CH_4 + O_2 \rightarrow H_2O + CO_2$

c $2CH_4 + 2O_2 \rightarrow 2H_2O + CO_2$

d $CH_4 + 2O_2 \rightarrow 2H_2O + CO_2$ [1]

10 Smart materials change their properties when they ...

a are used in computers

b detect an external stimulus

c are made into new materials

d become more volatile [1]

11 The most important property in a plastic used for toys is ...

a high tensile strength

b easily recycled

c non-toxic

d low density [1]

12 Kevlar's® impact resistance makes it useful for ...

a fire fighters' uniforms

b glass cutters' gloves

c high tensile ropes

d bullet-proof vests [1]

Physics P1a and P1b

1 A filament lamp is connected to a battery. The current in the lamp is measured with an ammeter. Electric current is measured in ...

a volts

b ohms

c amperes

d seconds [1]

2 In a laptop, a particular component is used to switch on a fan when the temperature gets too high. Which of the following components can you use for this task?

a Light-dependent resistor (LDR)

b Voltmeter

c Thermistor

d Ammeter [1]

3 The resistance of a filament lamp ...

a remains constant

b is infinite in one direction

c decreases when the current increases

d increases when the current increases [1]

4 The resistance of a component can be found using the equation:

$$resistance = \frac{voltage}{current}$$

	Voltage (V)	Current (A)	Resistance (Ω)
1	10	5.0	2.0
2	4.0	0.1	40
3	20	5.0	100

Which of the resistance readings are correct?

a 1 only

b 1 and 2

c 1 and 3

d 2 only   [1]

5 A filament lamp transforms 100 J of electrical energy into 6 J of light. The rest of the energy is wasted as heat. What is the efficiency of the lamp?

a 94%

b 106%

c 6%

d 100%   [1]

6 An electric kettle has a power rating of 1500 W. Over a period of a week, it is used for 2 hours. How many kilowatt-hours (kWh) of energy are transformed over this period of time?

a 3 kWh

b 0.75 kWh

c 3000 kWh

d 1.5 kWh   [1]

7 Here are some statements about a three-pin plug.

1 The fuse is connected to the neutral terminal.

2 The brown wire is connected to the live wire.

3 The blue wire is connected to the neutral terminal.

Which of the above statements are correct?

a 1 only

b 2 only

c 2 and 3

d 1 and 3   [1]

8 Ultrasound is a high frequency sound wave. Ultrasound can be used to image ...

a a foetus

b rain clouds

c plants

d a forged bank note   [1]

9 In a pond, an insect flapping its wings at a frequency of 200 Hz creates ripples on the surface of the water of wavelength 0.01 m. Use the equation below to find the speed of the ripples on the surface of the water.

speed = wavelength × frequency

a 200 m/s

b 0.01 m/s

c 2000 m/s

d 2.0 m/s   [1]

10 Which of the following waves is **not** a transverse wave?

a Sound

b Radio waves

c X-rays

d Microwaves   [1]

11 A fly-by probe can find the temperature of a planet by measuring the amount of ...

a infrared radiation emitted from the surface of the planet

b radio waves emitted by the planet

c X-rays absorbed by the planet

d infrared radiation emitted by the probe   [1]

12 In a TV programme, an astronomer mentions the following things:

1 The Universe contains billions of galaxies.

2 The galaxies are all moving away from each other.

3 The electromagnetic waves from the galaxies show red shift.

Which of the statements proves that the Universe is expanding?

a 1 only

b 2 only

c 1 and 2

d 2 and 3   [1]

Additional Biology

1 DNA codes for amino acids in a specific order to make specific proteins. Which organelle controls this?

a Nucleus

b Ribosomes

c Cytoplasm

d Cell wall [1]

2 In the lungs, gaseous exchange occurs. Tick the correct statement.

a Oxygenated blood enters the lungs and oxygenated blood leaves the lungs

b Oxygenated blood enters the lungs and deoxygenated blood leaves the lungs

c Blood leaving the lungs contains more carbon dioxide than blood entering the lungs

d Deoxygenated blood enters the lungs and oxygenated blood leaves the lungs [1]

3 A human body cell contains 23 pairs of chromosomes in its nucleus. A cell just having undergone mitosis would contain ...

a 23 single chromosomes

b 23 pairs of chromosomes

c 46 pairs of chromosomes

d 20 chromosomes [1]

4 The following statements describe the stages in cloning an animal. Put them in order.

a The egg is stimulated to divide and is then implanted into the surrogate mother where it further divides

b An egg cell is removed from the mother; its nucleus is removed

c The nucleus from the body cell is removed and placed into the egg cell

d A body cell is taken from the same organism [1]

5 It is a hot summer's day. Some tomato plants are growing in a greenhouse. Tick the limiting factor for photosynthesis in these plants.

a Light

b Carbon dioxide

c Temperature

d Space [1]

6 Nitrogen-fixing bacteria ...

a make up proteins

b convert nitrogen in the air into ammonia in the soil

c convert nitrates in the soil into nitrogen in the air

d break down plant and animal matter [1]

7 Animals compete for ...

a space, mates and light

b carbon dioxide, water and mates

c mates, territory and food

d nutrients, light and water [1]

8 Global warming is being caused by ...

a an increase in nitrogen

b an increase in acid rain

c an increase in the amount of CFCs used

d an increase in carbon dioxide [1]

Additional Chemistry

1 C_3H_6 is called ...

a propene

b propane

c butene

d butane [1]

2 Which property proves a compound contains C=C double bonds?

a It turns bromine water colourless

b It is produced during cracking

c It polymerises

d It has the formula C_nH_{2n} [1]

3 Which of the following makes a polymer thermosetting?

a Longer polymer chains

b Cross-links between chains

c Plasticiser between the chains

d Added preservatives [1]

4 A high melting point shows a metal has ...
  a delocalised electrons
  b strong metallic bonds
  c good conductivity
  d formed an alloy [1]

5 50% of bromine atoms have a mass of 79 and 50% have a mass of 81. What is bromine's relative atomic mass?
  a 79
  b 80
  c 81
  d 160 [1]

6 Potassium has 19 electrons so its electronic configuration is ...
  a 2.8.9
  b 2.8.1
  c 10.8.1
  d 2.8.8.1 [1]

7 Which of these combinations is not formed by carbon?
  a Four single bonds
  b Two double bonds
  c One triple and one single
  d One triple and one double [1]

8 In a nitrogen (N_2) molecule, each atom ...
  a gains three electrons
  b shares three pairs of electrons
  c forms one bond
  d loses three electrons [1]

9 The best description of the atoms in graphite is a ...
  a covalent molecule
  b giant covalent structure
  c giant ionic lattice
  d metallic crystal [1]

10 Reactions are faster when ...
  a reactants collide more frequently
  b reactants have less energy
  c dilute solutions are used
  d large lumps of solid are used [1]

11 Which of these would reduce the time magnesium takes to react with acid?
  a Adding more acid
  b Lowering the temperature
  c Using a more concentrated acid
  d Crushing the magnesium into a tight ball [1

12 The rate of reaction between Mg and HCl cannot be measured by ...
  a timing how long the Mg takes to disappear
  b measuring the time taken to make 10 cm^3 of hydrogen
  c recording the rate of pressure increase in a closed container
  d recording the colour change [1

Additional Physics

1 Which of the following is a safety feature in a car?
  a Steering wheel
  b Air bag
  c Exhaust
  d Tyres [1

2 The reaction time of a person is 0.5 seconds. What is the thinking distance when a car has a speed of 20 m/s?
  a 40 m
  b 4.0 m
  c 10 m
  d 100 m [1

3 The net force acting on an object is given by:
force = mass × acceleration
What is the acceleration of a car of mass 800 kg when the net force is 200 N?
  a 0.25 m/s
  b 0.25 m/s^2
  c 4.0 m/s
  d 4.0 m/s^2 [1

4 A monkey falls from a tree. Which of the following statements is true?

The monkey ...

a loses kinetic energy

b gains potential energy

c gains kinetic energy

d gains heat energy [1]

5 A moving car has kinetic energy. The kinetic energy of the car depends on its ...

a speed and weight

b speed and mass

c speed only

d mass only [1]

6 Albert Einstein created the Special Theory of Relativity. His ideas were the result of ...

a old ideas of physics

b real experiments

c thought experiment

d incorrect physics [1]

7 Which of the following statements is **not** correct for gamma rays?

Gamma rays ...

a are emitted from the nuclei of unstable atoms

b can be used to sterilise hospital equipment

c are used to heat up food in hospitals

d can be used to treat cancer patients [1]

8 Which of the following statements is **not** correct for isotopes?

a All isotopes belong to the same element

b All isotopes have the same number of protons

c ^1_1H and ^2_1H are examples of isotopes

d All isotopes have the same number of neutrons [1]

9 A detector is used to determine the activity of a short-lived isotope in a nuclear reactor. What happens to the activity of the source?

a It stays the same

b It decreases over a period of time

c It increases over a period of time

d It increases and then levels off [1]

10 Stars release their energy by ...

a fusion reactions

b fission reactions

c chemical reactions

d radioactivity [1]

11 Here are two statements about nuclear waste from power stations:

1 Burning can destroy the radioactivity of the waste.

2 Nuclear waste can remain radioactive for very long periods of time.

Which of the statements are correct?

a 1 only

b 2 only

c 1 and 2

d Neither [1]

12 Here are two statements about insulators:

1 Insulators can be charged by rubbing.

2 A positively charged insulator will repel a negatively charged insulator.

Which of these statements are correct?

a 1 only

b 2 only

c 1 and 2

d Neither [1]

GCSE Biology

1 We can use the bacteria *Agrobacterium tumefaciens* to transfer required genes into plant cells. One useful gene is made by *Bacillus theringiensis*. This gene ...

a enables the plant to grow more quickly

b changes the colour of the plant

c makes the plant poisonous to humans

d enables the plant to produce a toxin making it resistant to insects [1]

2 Tick two correct sentences.

a Taxol is found on the bark of Pacific yew trees

b Aspirin was one of the first treatments for cancer

c Aspirin is found on the bark of the Pacific yew tree

d Until the 1930s quinine was the only real treatment for malaria [1]

3 Humans are different from most animals because ...

a they do not have fur

b they are conscious of their actions

c they do not release pheromones

d they have a brain [1]

4 Tick the incorrect statement. Carnivores tend to ...

a spend less time eating compared to vertebrate herbivores

b have teeth specialised for grabbing and killing their prey

c lead a solitary life

d eat more than herbivores [1]

GCSE Chemistry

1 A brown liquid gives a white precipitate when nitric acid and silver nitrate are added, and a rust-coloured precipitate with sodium hydroxide. The best conclusion is that the liquid is:

a pure iron(III) chloride

b pure iron(III) sulphate

c a solution containing Fe^{3+} and Cl^- ions

d a solution containing Fe^{3+} and SO_4^{2-} ions [1]

2 Silver bromide can be prepared by precipitation. The correct equation for this is:

a $Ag^+_{(aq)} + NO_3^-{}_{(aq)} \rightarrow AgNO_{3(s)}$

b $Ag^+_{(aq)} + Br^-{}_{(aq)} \rightarrow AgBr_{(aq)}$

c $Br_{2(aq)} + 2AgI_{(aq)} \rightarrow 2AgBr_{(s)} + I_{2(aq)}$

d $Ag^+_{(aq)} + Br^-{}_{(aq)} \rightarrow AgBr_{(s)}$ [1]

3 It takes 40 cm^3 of HCl to neutralise 10 cm^3 of 1 M NaOH. What is the concentration of the HCl?

a 4 M

b 1 M

c 0.25 M

d 0.4 M [1]

4 The distinctive smell of pineapples is caused by an ester called ethyl ethanoate. Which of the following does **not** apply to esters?

a They have fruity smells

b They form when an alcohol reacts with an organic acid

c They fizz with carbonates

d They are useful solvents [1]

5 The Contact Process is carried out at 450 °C because:

a the reaction is most profitable at this temperature

b the yield is higher at higher temperatures

c the reaction is reversible

d it is too dangerous to increase the pressure [1]

6 When copper is purified by electrolysis:

a copper ions are attracted to the positive anode where they are oxidised to atoms

b copper ions are attracted to the negative cathode where they are oxidised to atoms

c copper ions are attracted to the positive anode where they are reduced to atoms

d copper ions are attracted to the negative cathode where they are reduced to atoms [1]

GCSE Physics

1 Here are two statements about temperature.

1 A temperature change of 1 K is equal to a temperature change of 1 °C.

2 100 °C is the same as 273 K.

Which of the statements are correct?

a 1 only

b 2 only

c 1 and 2

d Neither [1]

2 Which of the following statements is **not** correct for an ideal gas?

a For a fixed mass of gas at constant temperature, doubling its volume will double its pressure

b The volume of an ideal gas at 0 K is zero

c At 0 K, an ideal gas exerts no pressure

d The equation $\frac{PV}{T}$ = constant can be applied to an ideal gas [1]

3 Which of the following particles are fundamental particles?

a Neutrons

b Protons

c Atoms

d Electrons [1]

4 Which of the following statements describes thermionic emission?

a This is when electrons are accelerated in space

b This is when electrons are boiled off from hot metals

c This is when water boils

d This is when nuclei decay [1]

5 The intensity of a laser can be measured in:

a Joules (J)

b Joules second (J s)

c Watts (W)

d Watts per square metre (W m^{-2}) [1]

6 Here are two statements about radiation therapy.

1 Gamma rays or X-rays may be used to destroy cancerous cells.

2 Beta particles are used to destroy cancerous cells.

Which of the statements are correct?

a 1 only

b 2 only

c 1 and 2

d Neither [1]

7 Optical fibres are used in endoscopes. In an optical fibre, light is:

a refracted

b accelerated

c totally internally reflected

d turned into sound [1]

8 Which of the following is **not** true about a positron?

a It has a positive charge.

b It has the same mass as a proton.

c It is the antiparticle of an electron.

d It can combine with an electron to produce gamma rays. [1]

Model answers

1 The arrows on a food chain show ...

a the flow of energy

b wasted energy

c who eats who

d how much energy is lost

a is correct – it shows the flow of energy. c is not quite correct: even though you can figure out from a food chain what different organisms eat, the arrows actually represent the flow of energy. The wasted energy is not illustrated in a food chain nor the amount of energy lost.

2 Animals compete for ...

a light

b mates

c size

d soil

b is correct. Animals do not need light and soil to survive so they do not compete for these. Size is an inherited feature so is not competed for.

3 Genetic information is found ...

a in the nucleus of the cell

b in the cytoplasm of the cell

c outside the cell

d in the ribosomes

a is correct. Genetic information is found in the nucleus of cells.

4 An adaptation is a ...

a new feature

b feature that enables an organism to better survive its environment

c feature that means an organism is less likely to survive

d feature that allows organisms to live longer

b is correct – an adaptation is a feature that enables an organism to better survive its environment. For example, polar bears are adapted to live in the arctic because they have white fur meaning they camouflage well.

5 The motor neurone ...

a carries the message from the sensory to the relay nerve

b carries the message which senses the response

c carries the message which senses the stimulus

d carries the message which elicits the response

d is correct. The motor neurone links the effecter with the relay neurone. The motor neurone carries a message which causes the effecter to respond to a stimulus.

6 Oestrogen is at its highest level ...

a during fertilisation

b after the egg is released

c at the early part of the menstrual cycle

d throughout the menstrual period

c is correct. Oestrogen causes the lining of the uterus to thicken during the early part of the menstrual cycle.

7 Antibodies are produced by white blood cells to defend the body against pathogens. An antibody must be ...

a specific to the antibody on the surface of the pathogen

b different to the antigen on the surface of the body cell

c destroyed by the white blood cell

d specific to the antigen on the surface of the pathogen

d is correct. Antigens are proteins found on the surface of pathogens. White blood cells produce specific antibodies which match the antigen exactly.

8 A definition of a drug is ...

a an illegal substance

b anything that makes you happy

c anything that is addictive

d any substance that modifies the body if taken

d is correct. A drug will not necessarily be illegal. A drug is any substance which, when taken into the body, modifies it chemically.

Chemistry C1a and C1b

1 Which of the following could be potassium chlorate?

a KCl

b KCL

c $KClO_4$

d $KCLO_4$

They all start with a K so they all contain potassium. b and d are completely wrong because the symbol for chlorine is Cl not CL. When there are two letters in a symbol, the second one is never a capital. a is incorrect because non-metals change their ending to -ide when they react with metals e.g. chlorine becomes chloride. c is the correct answer. The ending -ate shows that the compound contains oxygen as well as potassium and chlorine.

2 If sodium hydroxide is added to a solution of a copper compound:

a it turns blue

b the copper dissolves

c the copper is displaced

d an insoluble hydroxide forms

a is definitely wrong. Solutions of copper compounds only turn a deep blue colour when ammonium hydroxide is added. The copper is already dissolved, so b cannot be right. c is not true because copper atoms are not displaced from the solution. The reaction forms an insoluble solid when the two solutions are mixed, so it is a precipitation reaction. The precipitate is insoluble copper hydroxide, so d is the correct answer.

3 Silver iodide is insoluble so it cannot be made by neutralising an acid.

Choose two solutions that would react to make it:

a silver and iodine

b silver nitrate and lead iodide

c silver nitrate and potassium iodide

d silver nitrate and iodine

The silver iodide needs to be made by precipitation. A precipitate is formed by mixing solutions, so the silver and the iodide both need to be in soluble compounds. A nitrate is a good choice because all nitrates are soluble. a and d are wrong because iodide is needed, not iodine. b is no good because lead iodide is insoluble. c is the correct choice and it is worth remembering that sodium and potassium compounds always dissolve.

4 Downward delivery is used to collect CO_2 because it:

a turns limewater milky

b puts out fires

c is less dense than air

d is more dense than air

a and b are both true, but they do not explain why downward delivery is used to collect CO_2. In downward delivery the gas being collected has to push the air in the container upwards. To do this it has to be more dense than air, so c is wrong and d is correct. It is worth remembering that gases less dense than air are collected by upward delivery. If these gases do not need to be dry, they can also be collected over water if they are not too soluble.

5 It is dangerous to restrict the oxygen supply to a gas burner because:

a more energy is released

b carbon dioxide is formed

c carbon monoxide may form

d complete combustion takes longer

Gas burners work with hydrocarbon fuels like methane. When these burn, they normally make carbon dioxide and water, so b is wrong. When the oxygen supply is restricted, water forms but carbon monoxide can be made instead of carbon dioxide. Complete combustion does not take place so d is wrong. This also means that less energy is released, so a is wrong. The carbon monoxide is dangerous because it reduces the blood's ability to transport oxygen, so c is the correct answer.

6 The % of children with asthma is higher where SO_2 concentrations are higher. This indicates that:

a SO_2 causes asthma

b SO_2 might cause asthma

c SO_2 does not cause asthma

d SO_2 relieves the symptoms of asthma

The link between SO_2 levels and asthma is known as a correlation. A correlation does not imply that one thing causes another, but rules out c and d. We cannot be certain that a is true without more information, so b is the correct answer.

7 Kevlar's® impact resistance makes it useful for:

a fire fighters' uniforms

b glass cutters' gloves

c high tensile ropes

d bullet-proof vests

Kevlar has a range of useful properties and is used for all these things. Fire fighters' uniforms rely on its fire resistance, so a is wrong. Glass cutters' gloves rely on its cut resistance, so b is wrong. High tensile ropes rely on its tensile strength, so c is wrong and d is the correct answer.

8 Emulsifiers:

a make oil dissolve in water

b can be hydrophilic or hydrophobic

c suspend droplets of one liquid in another

d make liquids separate into two layers

Emulsifiers stop liquids separating into two layers, so d is wrong. They cannot make oil dissolve in water, so a is wrong. c is the correct answer. The emulsifiers are able to do this because they have hydrophilic ends, which are attracted to oil, and hydrophobic ends, which are attracted to water. So b is also wrong.

Physics P1a and P1b

1 A filament lamp is connected to a cell. The brightness of this lamp can be increased by ...

a connecting a resistor in series with the lamp

b connecting another lamp in series

c connecting another lamp in parallel

d using two cells

Connecting a resistor will increase the resistance of the circuit. Hence the current in the lamp will decrease. This will make the lamp dimmer, not brighter. Hence a is incorrect.

Connecting another lamp will also increase the

resistance and hence decrease the current. As in a, the lamp will be dimmer. Hence b is incorrect.

The brightness of the lamp will not change because the voltage (potential difference) across the lamp does not change. Hence c is incorrect.

Using another cell will increase the voltage across the lamp and hence the brightness of the lamp will increase. The correct answer is d.

2 It is dangerous to dispose of chemical batteries because they ...

a have toxic metals

b heat the environment

c cause acid rain

d burn the ground

Batteries contain metals like cadmium and lead. These are very dangerous because they are toxic. So the answer is a. The other answers just do not make sense. For example, burning fossil fuels causes acid rain and has nothing at all to do with the battery. A battery has chemicals inside and does not burn coal!

3 A heater has a power rating of 2.4 kW. What is this power in watts?

a 0.0024 watts

b 2 400 000 watts

c 2400 watts

d 2.4 watts

It is important to understand that the 'k' here refers to 'kilo', which is a factor of 1000. Hence, 2.4 kW is the same as 2400 W. The correct answer is c. You will get a as the answer if you think that the 'k' is equal to 1/1000. In b, you have gone too far by assuming that 'k' is equal to one million. The answer d is just too silly because it assumes that you have not seen the 'k' in front of the watts (W) symbol.

4 Which of the following equations can be used to calculate the power of an electric appliance?

a power = voltage/current

b power = voltage × current

c power = current/voltage

d power = voltage + current

It is very important that we remember any equations correctly. The correct answer is b. The answers a and c are simply incorrect. The answer d makes no sense because here you are adding two different quantities to produce power.

5 What do X-rays and gamma rays have in common?

a Both have the same wavelength

b Both have the same frequency

c Both are longitudinal waves

d Both can travel through a vacuum

X-rays and gamma rays are both electromagnetic waves. All electromagnetic waves are transverse waves, can travel through a vacuum but they all have different frequencies or wavelengths. Hence the correct answer must be d.

6 The speed of a wave is given by the equation:

speed = frequency × wavelength

	Speed (ms)	Frequency (Hz)	Wavelength (m)
1	240	120	2.0
2	234	680	2.0
3	4.0	10	0.4

Which of the speed values are correct?

a 1 only

b 3 only

c 2 and 3 only

d 1 and 3 only

It is best to use the given equation to find the speed of the wave in each case.

For speed 1, we get:
speed = 120 × 2.0 = 240 m/s
Hence statement 1 is correct.

For speed 2, we get:
speed = 680 × 2.0 = 1360 m/s
Hence statement 2 is incorrect.

For speed 3, we get:
speed = 10 × 0.4 = 4.0 m/s
Hence statement 3 is correct.

Only 1 and 3 are correct, so the answer must be D.

7 We know that the Universe is expanding because ...

a the planets move round the Sun

b the Solar System is expanding

c the galaxies are moving further apart

d universes are moving further apart

There is one Universe; hence d cannot be the answer. The planets moving round the Sun have nothing to do with the expansion of the Universe. b is going to be a popular distracter because it contains the word 'expansion'. b cannot be the answer because there is no reference to galaxies. c is the correct answer. As the space of the Universe expands, it carries with it all the galaxies.

8 Ultrasound is a high frequency sound wave. What is meant by 'frequency'?

a It is the distance between two adjacent peaks

b It is the maximum displacement of a particle from its rest position

c The number of waves passing through a point per second

d It is measured in metres

a is really defining the wavelength of a wave, so this cannot be the answer. b is defining the amplitude of a wave. c is the correct answer because this is how frequency is defined. Frequency is measured in hertz (Hz), so d is way off the mark!

Additional Biology

1 DNA is found in the nucleus of living cells. A section of DNA which codes for a specific protein is known as a ...

a characteristic

b protein

c chromosome

d gene

d is correct. Chromosomes are found in the nucleus of living cells; they are made up of a chemical called DNA. Sections of chromosomes which code for a certain characteristic are known as genes.

2 Tick the equation that shows aerobic respiration.

a oxygen + glucose → carbon + water + energy
dioxide

b carbon dioxide + glucose → lactic acid + energy

c glucose → lactic acid + energy

d oxygen + glucose + water → carbon + energy
dioxide

a is correct. Oxygen and glucose are taken into the body cells where respiration occurs, making energy, carbon dioxide and water. c is anaerobic respiration.

3 Tick the two types of cells that would undergo mitosis.

a Muscle cell

b Skin cell

c Sperm cell

d Egg cell

a and b are correct. Body cells undergo mitosis, sex cells undergo meiosis.

4 Tick the two correct statements.

a Cells can only divide a certain number of times; this is known as the Hayflick limit

b We can use normal body cells to grow any other body part

c Cancer cells have a Hayflick limit

d Stem cells have not yet specialised into a certain cell type

a and d are correct. Body cells are specialised to a certain body part and cannot be used to grow any part. Stem cells on the other hand are not specialised and can be stimulated to grow into anything. Cancer cells have no limit to the number of times they will divide.

5 The correct equation for photosynthesis is ...

a glucose + carbon → oxygen + water + energy
      dioxide

b oxygen + carbon dioxide → water + glucose

c carbon dioxide + glucose → oxygen + water

d carbon dioxide + water → oxygen + glucose

d is correct – plants take water and carbon dioxide and make glucose and oxygen.

6 Which human activity is causing an increase in the amount of carbon dioxide in the Earth's atmosphere?

a Use of CFCs

b Burning fossil fuels

c Photosynthesis

d Using wind power

b is correct – when fossil fuels are burnt, carbon dioxide and water are released. Photosynthesis removes carbon dioxide from the atmosphere and wind power is a clean energy resource.

7 Use the following food chain to answer the next question.

grain ⟶ mouse ⟶ owl

Which statement is correct?

a As the population of owls increases, the population of mice decreases

b If the mouse population is reduced due to disease, the owl population will increase

c If more grain is made available, the mouse population will decrease

d If the mouse population increases, the owl population could also increase

a is correct. Mice are the prey for owls so if there are more owls, more mice will be eaten to support the larger owl population.

8 Which graph would show the average temperature of Earth over the past 100 years?

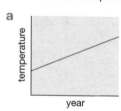

a is correct. It shows the line of best fit for average global temperature. The global temperature is gradually rising.

Additional Chemistry

1 What is the theoretical yield of CaO from 100 g of $CaCO_3$?

$CaCO_3 \rightarrow CaO + CO_2$

a 100 g

b 56 g

c 44 g

d 28 g

100 g corresponds to 1 formula mass of $CaCO_3$. You work this out by adding the mass numbers: Ca = 40, C = 12 and 3 x O = 16. According to the equation, the theoretical yield is 1 formula mass of CaO which is 56g.

2 If 100 g of CaO makes 14 g of a CaO. What is the percentage yield?

a 86%

b 50%

c 25%

d 14%

The theoretical yield of CaO is 56 g, so 56 g represents a 100% yield. 14 g is $\frac{1}{4}$ of this, which is 25%.

3 Which of these is not a group 7 property?

a Forms 1+ ions

b Has 7 electrons in its outer shell

c Reacts by gaining 1 electron

d Forms 1- ions

Metals have three electrons or fewer in their outer shells and form positive ions by losing them. Chlorine is not a metal, so a is wrong. b is true because chlorine is in group 7. It reacts by gaining 1 electron, so c is true. Electrons are negative so d is also true.

4 Organic farming is thought to be more sustainable than conventional farming because:

a it maintains soil structure

b it gives lower yields

c it is more labour intensive

d it uses fewer chemicals

Organic farming does use fewer chemicals. The lack of artificial fertiliser does make yields lower. More labour intensive work like weeding is needed because no herbicides are used. However, these effects are balanced by the reduced pollution and lower energy costs. They are not a threat to sustainability. a is the correct answer.

Additional Physics

1 The braking distance of a car will increase ...

a when going up a hill

b when travelling along an icy road

c when the car is travelling slowly

d when being towed

The braking distance of a car depends on the speed of the car, the condition of the road and the condition of the car's tyres and brakes. Going up a hill will increase the potential energy of the car, hence the braking distance will decrease. Hence, a is not the answer. On an icy road, the friction between the tyres and the road will increase, hence the braking distance will also increase. b must be the answer. A car travelling slowly will not travel too far when the brakes are applied, therefore c cannot be the correct answer. d just cannot be the correct answer because the car is being towed by another vehicle.

2 The kinetic energy of an object is given by:

kinetic energy = $\frac{1}{2} \times$ mass $\times$ speed^2

Two identical cars are travelling at different speeds. Car X is travelling at 10 m/s and car Y is travelling at 20 m/s. Which of the following statements is **not** true?

a The cars have different kinetic energies

b Car Y has greater kinetic energy

c Car Y has twice the kinetic energy of car X

d Car Y has four times greater kinetic energy than car X

The cars are identical, so they must have the same mass. Since the cars are travelling at different speeds, they must have different kinetic energies. Hence a is a correct statement. The faster travelling car Y will have a greater kinetic energy. Hence b is also a correct statement. Car Y has twice the speed of car X. Since the kinetic energy depends on speed^2, this means that the kinetic energy will increase by a factor of four and not two. The statement c is incorrect. Since the question wants you to find the incorrect statement, the correct letter for the answer would be c.

3 How many neutrons does the isotope $^{235}_{92}U$ have?

a 235

b 92

c 327

d 143

The nucleon or mass number is 235 and the proton or atomic number is 92. There are 235 neutrons and protons altogether. The number of neutrons must therefore be (235 − 92) = 143. Hence the correct answer is d.

4 The half-life of an isotope is 10 years. Here are some possible statements concerning a radioactive source with this isotope:

1 After a time interval of 20 years, $\frac{1}{4}$ of the active nuclei will be left.

2 The activity of the source will decrease with time.

Which of the statement(s) is/are correct?

a 1 only

b 2 only

c 1 and 2

d None

The activity of a source is equal to the number of nuclei decaying per unit time. As the number of active nuclei decreases, the activity of the source will decrease. Statement 2 is therefore correct. After each half-life, the number of active nuclei will halve. After 20 years, which is two half-lives, the number of active nuclei will be one quarter of the original number of nuclei. Statement 1 is also correct. Hence the correct answer will be c.

GCSE Biology

1 Tick the incorrect statement.

a Bacteria can be used to produce yoghurt from milk

b Enzymes produced by the bacteria *Aspergillus niger* are used to make the middle of sweets softer

c Enzymes in washing powder break down stains on clothes

d Enzymes are man-made catalysts

d is incorrect. Enzymes are biological catalysts. They work at a specific temperature and pH, and speed up reactions.

2 With which two medical conditions is obesity linked?

a Coronary heart disease

b Tuberculosis

c High blood pressure

d Low blood pressure

a and c are correct. Fatty deposits line and harden the arteries. This results in higher blood pressure and also heart disease.

3 Which of the following is **not** a form of communication between animals?

a Sounds

b Body posture

c Lifestyle

d Facial expressions

c is incorrect. Lifestyle is not how animals communicate. They communicate via sound, facial expressions, chemicals (pheromones) and posture.

4 Tick the two correct statements. Vertebrate herbivores tend to ...

a graze in large groups

b lead a solitary life

c have special teeth that are evolved to cope with constant chewing and grinding

d have sharp teeth for catching

a and c are correct. Vertebrate herbivores tend to graze in large numbers for protection. They also have specialised teeth to cope with the high wear and tear from constant grinding. b is incorrect − carnivores are more likely to lead a solitary life.

GCSE Chemistry

1 A solution contains 80 grams dm^{-3} of NaOH. What is the concentration in $mol\ dm^{-3}$?

a 2

b 1

c 40

d 20

A 1 $mol\ dm^{-3}$ solution contains 1 mole of solute per dm^3, which is the relative formula mass of the solute in grams. For NaOH this is 40 g (Na = 23, O = 16, H = 1). 80 g is twice as much as this so the concentration is 2 $mol\ dm^{-3}$, making a the correct answer. You can get the same answer by using the equation: grams dm^{-3} = $mol\ dm^{-3}$ x M_r.

2 Which of the following is not true of esters?

a Made from an acid and an alcohol

b Fruity smell

c The main component of soap

d Good solvent

Organic acids and alcohols react to make esters. These have a characteristic fruity smell and make good solvents, so a, b and d are true. They are used to add scent to soap, but the main components of soap are organic acids. c is the correct answer because it is the only one that is untrue.

GCSE Physics

1 What is meant by thermionic emission?

a Electrons emitted by a radioactive source

b Electrons emitted from cold metals

c Electrons 'boiled off' a hot filament

d Light emitted by a filament lamp

When a metal is heated, the electrons close to the surface of the metal have sufficient kinetic energy to escape the surface of the metal. This is thermionic emission. c must be the correct answer. a makes reference to beta-emission. Although electrons are mentioned, this is not thermionic emission. Electrons will not escape easily from cold metals, therefore b is incorrect. The statement b makes reference to light and this cannot be anything to do with thermionic emission.

2 Which of the following particles may be regarded as fundamental?

a Protons and neutrons

b Electrons and protons

c Quarks and neutrons

d Quarks and electrons

Neutrons and protons are not fundamental particles because they are made up of quarks. Hence, a, b or c cannot be the correct answer. The quarks are definitely fundamental particles because they cannot be sub-divided. Electrons too are fundamental particles. The answer is confirmed to be d.

Answers

NOW TRY THIS ANSWERS

Environment 1 – pages 6 and 7
a eat, active, live, variation, food
b i Gets its energy by consuming other organisms
  ii Produces its own food by photosynthesis
  iii Shows the flow of energy through a community
  iv The mass of all living matter in an area
  v The organism which is consumed by a predator
c Conventional, organic, conventional, organic, conventional, organic
d decreases, increase, increase, decrease

Environment 2 – pages 8 and 9
a Factors ticked: Birth rates, Medicine, Average life expectancy, Improved sanitation, Better understanding of disease, Diet
b adaptations, beneath, gradual, extinct
c i Organisms best adapted to their environment will survive and pass on their genes
  ii Found in the nucleus of body cells, these determine your inherited characteristics
  iii Characteristic that enables an organism to better survive in its environment
  iv Changing in order to become better adapted to an environment
  v Scientist who produced the theory of evolution
d Natural, selective, genetic, natural, genetic, selective and genetic

Genes 1 – pages 10 and 11
a chromosomes, genes, characteristic, DNA, inherited
b Asexual, sexual, asexual, asexual, sexual, asexual, sexual
c

	B	b
B	BB	Bb
b	Bb	bb

1:4 (25%), 3:4 (75%), brown, brown
d characteristics, mutation, haemophilia, cystic fibrosis

Genes 2 – pages 12 and 13
a Inherited, environmental, inherited and environmental, environmental, environmental, inherited
b 6, 3, 4, 1, 5, 2
c genes, cured, cystic fibrosis, haemophilia, intelligence, looks
d Answers will vary

Electrical and chemical symbols 1 – pages 14 and 15
a Ticked: Alcohol intake, Caffeine intake, Age of subject, Sex of subject
b shivering, hairs, increasing, ducking, survival
c i Found in the skin
  ii Carries hormones around the body
  iii Example: light
  iv Centre in the brain which monitors body temperature
  v Example: blood glucose levels
  vi Keeping the internal conditions of the body constant
d carbon dioxide, urea, oxygen/glucose, hormones

Electrical and chemical symbols 2 – pages 16 and 17
a Hormone, reflex, hormone, reflex, reflex
b fertilisation, hormone, egg, sperm, uterus
c

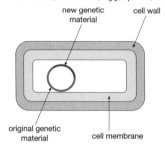

new genetic material — cell wall — original genetic material — cell membrane

Use, misuse and abuse 1 – pages 18 and 19
a Bacteria, virus, fungi, virus, fungi, bacteria, fungi, virus, fungi, bacteria
b microbes, mouth, eyes, ears, disease, vector, direct, air, food
c i White blood cells engulf bacteria
  ii Small hairs found in the trachea, prevent microbes from entering the lungs
  iii An enzyme found in tears
  iv Any organism not visible to the naked eye
  v Cell with a very large nucleus
  vi Any disease-causing organism
d i White blood cells engulf and ingest bacteria
  ii Released by white blood cells as a response to a pathogen
  iii Cells found in the blood that play a large part in immunity
  iv These cells survive much longer than other white blood cells, and are able to elicit a fast response to a pathogen that has been met before
  v Proteins found on the surface of foreign cells
  vi The ability to resist a certain disease

Use, misuse and abuse 2 – pages 20 and 21
a pathogen, disease, immune, pathogen, antibody
b Symptoms of TB: coughing up blood, fatigue, scarring of lungs, night sweats, fever, weight loss
c False, true, false, false, true
d drug, drug, depressant, blocks, reaction, liver

All about atoms – pages 22 and 23
a Proton: positive, in the nucleus, number different for every element; Neutron: in the nucleus, neutral; Electron: negative, around the outside of an atom
b Co, NO, H_2O, $CuSO_4$, CuS
c iron oxide, copper sulphide, potassium chloride, $2HBr$, $2Mg + O_2 \rightarrow 2MgO$, $2H_2 + O_2 \rightarrow 2H_2O$

d Exothermic, exothermic, endothermic, exothermic

The elements – pages 24 and 25
a i Cu ii Na iii Cu iv Fe(II) v Li vi Fe(III) vi Ca viii K
b Chlorine: displaces bromine, green gas, most reactive, makes white salts; Bromine: has a brown solution, brown liquid, makes white salts, only displaces iodine; Iodine: grey solid, has a brown solution, makes white salts
c i, ii, v, vi
d Transition block, group 7, group 7, noble gas, transition block, group 1, transition block, noble gas

Chemicals everywhere – pages 26 and 27
a i making fertilisers
  ii dissolving grease
  iii making bread rise and making drinks fizz
  iv dissolving limescale
  v flavouring lemonade
  vi providing energy,
  vii flavouring foods
b HCl, KOH, $NaNO_3$, CuO, $CaCO_3$, all three
c lead iodide + sodium nitrate, barium sulphate + sodium nitrate, silver iodide + sodium nitrate, AgI + NaNO3, silver chloride + sodium nitrate, AgCl + $NaNO_3$
d Reduced using electricity: Al, Na; Reduced by carbon: Fe, Pb; Found as element: Ag, Au

Everyday chemicals – pages 28 and 29
a i releases CO_2 if heated, contains CO_3
  ii breaks up compounds
  iii addition of water
  iv contains HCO_3, releases CO_2 if heated, found in baking powder
  v loss of water, makes food brown
  vi decomposition product
  vi dehydrated starch
b H_2, H_2, O_2, CO_2, H_2, Cl_2, CO_2, NH_3, NH_3, both Cl_2 and NH_3
c True, False, False, True, False, True, False, False, True
d Hydrogen: difficult to store, used in fuel cells, pollution-free, releases H_2O only; Biodiesel: releases CO_2 and H_2O, greenhouse-neutral, made from plant oils; Ethanol: releases CO_2 and H_2O, greenhouse-neutral, made from sugar; Fossil fuel: releases CO_2 and H_2O

Fuels and air quality – pages 30 and 31
a Small, Small, Small, Large, Large, Large, Small, Small, Large
b carbon dioxide + water, CO_2, carbon monoxide + water, CO, carbon + water, C
c increasing, decrease, cannot
d Increase, increase, decrease, increase, increase, decrease, decrease, increase, increase

New materials – pages 32 and 33
a i PTFE ii shape memory alloy
  iii thinsulate iv lycra v carbon fibre

b
i parachute ropes
ii bullet-proof vests, safety helmets
iii fire fighter's uniform, bullet-proof vests, safety helmets
iv run-flat tyres, fire fighter's uniform
v glass workers' gloves
c i, iii, iv, vi
d
i a reaction with oxygen
ii disperse one liquid in another
iii stops emulsions separating
iv water loving
v lets microbes grow
vi water hating
vii cannot be mixed

Measuring electricity – pages 34 and 35

a current, 0.20 A, voltmeter, 10 V

b

V		L	A	M	P
O					
L		G			
T	N	E	R	R	U	C
A			A		
G				P	
E					H
R	E	S	I	S	T	O	R

c Second, third and fourth statements ticked

Producing electricity – pages 36 and 37

a **i** electrical **ii** does not **iii** 30 **iv** toxic
b
i A dynamo produces an alternating current
ii To induce a current in a wire you have to move it in a magnetic field
iii In a dynamo the magnet rotates near a stationary coil
iv Rotating the coil of a dynamo faster increases the frequency of the current
c **i** small **ii** energy **iii** fast **iv** electromagnets

Being in charge – pages 38 and 39

a steam, coil, magnet, coil, Grid
b Third statement ticked
c

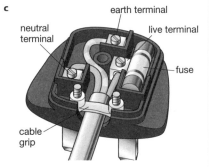

earth terminal
neutral terminal
live terminal
fuse
cable grip

It's all about waves – pages 40 and 41

a
i m/s
ii m
iii Hz
iv earthquake
b True, true, false, true, false
c electromagnetic, vacuum, speed, shorter, cancer

Uses and dangers of waves – pages 42 and 43

a Radio waves – no damage; Microwaves – heating; Infrared radiation – heating; Ultraviolet radiation – sunburn, cancer; X-rays – cancer; Gamma rays – cancer

b

I			R				
	R		E	N	O	B
A		I		F			
B		S	L		N	N
S			E		A	O
O			C		C	T
R	F	O	E	T	U	S	E
B						S

c True, false, true, true

Space – pages 44 and 45

a True, false, true, false, false
b
i the same
ii 20 N
iii 8.0 N,
c Interplanetary space has no air; Astronauts have to exercise regularly to prevent muscle wastage; Landers and fly-by probes are unmanned spacecrafts; There is a small chance of extraterrestrial life on other planets in the Universe

Stars and the Universe – pages 46 and 47

a **i** gravity **ii** fusion **iii** billions of years
b For a star like our Sun: star, red giant, planetary nebula, white dwarf, black dwarf; For a star more massive than our Sun: star, red supergiant, supernova, black hole
c The Universe began from an event known as the Big Bang; All galaxies are moving away from each other because the Universe is expanding; The light from all galaxies shows red shift; If there is not enough matter in the Universe it will expand forever and get cooler

Inside living cells – pages 48 and 49

a four, three, one, amino acids, mutation, amino acid, amino acids, protein
b Trait: A characteristic; G: Guanine; Base: T, A, G and C are examples of these; C: Cytosine; DNA: Deoxyribonucleic acid; A: Adenine; T: Thymine; Gene: A section of DNA that codes for a certain characteristic
c Amino acid: Each codon codes for an amino acid; Ribosome: This is an organelle that uses RNA to make proteins; Three: The number of bases in each codon; Base: For example, A, T, G and C; Codon: Three bases make up a codon; Protein: Many amino acids join together to make this; mRNA: A single-stranded copy of DNA
d Aerobic, anaerobic, aerobic, anaerobic, anaerobic, anaerobic, anaerobic

Divide and develop 1 – pages 50 and 51

a grow, repair, mitosis, somatic, identical, chromosomes
b Mitosis, meiosis, mitosis, both, mitosis, meiosis, mitosis, meiosis, meiosis, meiosis
c
i Determine the hayflick limit
ii This type of cell has no hayflick limit
iii The limit to the number of divisions a cell can undergo
iv This type of cell is not yet specialised
v Mitosis or meiosis
d Continuous, discrete, continuous, discrete, continuous, discrete, discrete, discrete, continuous

Divide and develop 2 – pages 52 and 53

a light, water, nutrients, space, dispersed, air, fruit, digestive
b selective breeding, features, characteristics, bred, dogs, cows, horses, wheat
c False, true, true, true, true, true, true, false
d Cancer, cystic fibrosis, Huntington's chorea, Downs syndrome, Sickle cell anaemia, Diabetes

Energy flow 1 – pages 54 and 55

a

nucleus
cell membrane
cytoplasm

b
i Gas produced by photosynthesis and used for respiration
ii Organelle found in plant cells where photosynthesis occurs
iii Type of energy provided by the Sun, needed for photosynthesis
iv Gas which is a reactant of photosynthesis
v Product of photosynthesis, often stored in the fruit
vi Green pigment
vii Liquid reactant of photosynthesis
c True, false, true, false, true, false
d Burning fossil fuels, respiration, decomposing, burning wood

Energy flow 2 – pages 56 and 57

a
i Convert nitrogen in the air into nitrates in the soil
ii Break down dead plant and animal matter, returning nitrates to the soil
iii Element important in amino acids
iv Make up protein
v Convert nitrates in the soil into nitrogen in the air
b 6, 8, 7, 3, 5, 2, 1, 4
c Controlling the temperature, Keeping animals indoors, Restricting animal movement, Feeding animals antibiotics, Protecting animals from predators
d Ticked activities: Burning fossil fuels, Deforestation, Intensive farming

Interdependence – pages 58 and 59

a Habitat: A part of the environment in which an organism lives; Interdependence: The population size of a predator and prey depend on each other; Predator: An animal that lives by feeding off other animals; Population: A group of interbreeding species in a habitat
b carbon dioxide, nitrates, carbon dioxide, burning fossil fuels, exponentially, CFCs, global warming
c
i Has resulted in the destruction of much of the countryside
ii An area containing a great deal of biodiversity – areas where conservation efforts are concentrated
iii Affected greatly by deforestation
iv Number and variety of species in an area
v Human intervention to preserve, protect and manage the environment

d reused; recycled; glass, metal, paper; land fills, burning

Carbon compounds – pages 60 and 61

a ethane, methane, propene, propane, butane

b **i** saturated hydrocarbon
ii contains no C=C bonds
iii decomposition
iv unsaturated hydrocarbon
v compound of H and C,
vi contains a C=C bond, C_2H_5OH

c polystyrene, polyethene, polybutene, polyethene, polypropene, PVC, PVC

d iv, vi

Making changes – pages 62 and 63

a A, A, A, B, B, A, B

b compounds, mixtures, two, four, target, clinical

c 56, 30, 46, 74, $H_2 + I_2 \rightarrow 2HI$, $2Mg + O_2 \rightarrow 2MgO$, $2Na + Cl_2 \rightarrow 2NaCl$, $Mg + 2HCl \rightarrow MgCl_2 + H_2$, $Mg + 2HNO_3 \rightarrow Mg(NO_3)_2 + H_2$, $C_2H_4 + 3O_2 \rightarrow 2CO_2 + 2H_2O$, $C_3H_8 + 5O_2 \rightarrow 3CO_2 + 4H_2O$

d MgO, CaO, CO_2, CH_4, SO_2, NH_3, C_2H_6, H_2O

Elements – pages 64 and 65

a Strong bonds cause: high tensile strength, high melting points; Delocalised electrons: cause conductivity, are negatively charged; Alloys: are mixtures of metals, have useful properties; Flexible bonds: make metals malleable

b Protons, protons and neutrons, protons, isotopes, protons, protons

c B (2,1), N (2,5), Ne (2,8) Mg (2,8,2), Cl (2,8,7), Ca (2,8,8,2), Al (2,8,3), P (2,8,5)

d $Mg^{2+}_{(2,8)}$; $Cl^-_{(2,8,8)}$; – 1 electron $\rightarrow Na^+_{(2,8)}$; – 3 electrons $\rightarrow Al^{3+}_{(2,8)}$; + 2 electrons $\rightarrow O^{2-}_{(2,8)}$; – 1 electron $\rightarrow K^+_{(2,8)}$; – 2 electrons $\rightarrow Ca^{2+}_{(2,8,8)}$; $F_{(2,7)}$ + 1 electron $\rightarrow F^-_{(2,8)}$; $S_{(2,8,6)}$ + 2 electrons $\rightarrow S^{2-}_{(2,8,8)}$; $Li_{(2,1)}$ – 1 electron $\rightarrow Li^+_{(2)}$

Explaining properties – pages 66 and 67

a K_2O, $CaCl_2$, NaI, $Ca(OH)_2$, CaO, K_2S, $Ca(NO_3)_2$, Na_2SO_4, MgI_2, Li_2O

b 1, 8, 7, 7, 7, 1, 8

c hydrogen, H_2, potassium hydroxide + hydrogen, $2KOH + H_2$, bromine, Br_2, potassium bromide + iodine, $2KBr + I_2$

d ions, electrodes, negative/cathode, electrons, move, positive/anode, lose, molecules, decomposed

Molecules and metals – pages 68 and 69

a H–H, N≡N, Cl–Cl, O=O, H–Cl, H–O–H, O=C=O

b Graphite, diamond, graphite, graphite, diamond, graphite, diamond, diamond, diamond, graphite, graphite

c Check diagrams with your teacher

d Molecular, metallic, molecular, giant covalent, metallic, giant covalent, molecular, metallic

Using chemistry – pages 70 and 71

a therapies, clinical, lab, animals, volunteers, existing, placebo

b A, C, D, B

c **i** $H_{2(g)} + I2_{(s)} \rightarrow 2HI_{(g)}$
ii $2H_{2(g)} + O_{2(g)} \rightarrow 2H_2O_{(l)}$
iii $Fe_{(s)} + S_{(s)} \rightarrow FeS_{(s)}$
iv $N_{2(g)} + 3H_{2(g)} \rightarrow 2NH_{3(g)}$

d Exothermic, exothermic, exothermic, endothermic, exothermic, endothermic, exothermic, exothermic, exothermic, endothermic, endothermic

Reaction rates – pages 72 and 73

a Slower, faster, slower, slower, faster, slower, neither

b Light, temperature, pH, pressure or mass, pressure, light, temperature, pH or pressure

c More, less, more, more, neither

d N_2: nitrogen, molecule, found in air; NH_3: molecule, ammonia; NH_4^+: ammonium, ion, in fertiliser; NO_3^-: ion, nitrate, in fertiliser

Understanding motion – pages 74 and 75

a **i** scalar **ii** vector **iii** acceleration
iv acceleration

b False, true, true, false

c

N		H	T	R	A	E
	E					
		W				
S		E	T	F		
S		I	O			
A		G	R	N		
M		H	C		S	
		T	E			

Falling and collisions – pages 76 and 77

a True, true, false, true

b 6.0, braking, thinking, stopping

c **i** deaths
ii in control,

Work and energy – pages 78 and 79

a No work is done on an object if it remains stationary; Power is measured in watts; Work is measured in joules; In the equation: $W = F \times s$, W is work done

b False, true, false, false

c Third and fourth statements ticked

Einstein – pages 80 and 81

a

	S	P	E	C	I	A	L	
		L	I	G	H	T		
						H		
N	I	E	T	S	N	I	E	
			E			O		
			S			R		
						T	Y	
	T	H	G	U	O	H	T	

b First, second, third, fifth and sixth statements ticked

Radioactivity – pages 82 and 83

a **i** positive **ii** proton **iii** element

b cancer, granite, radon, low, naturally

c First and second statements ticked

d **i** Gamma rays kill off bacteria
ii Smoke detectors use americium
iii The isotope of americium has 241 nucleons
iv The carbon-14 isotope is radioactive,
v Age of rocks can be found from the decay of uranium nuclei

a X-rays: i, iii, iv, v
Gamma rays: i, ii, iv

b

			I			
	G		O		S	
	A		N		T	
	M		I		E	
X	M		S	C	R	
T	R	A	C	E	R	I
	A				L	
	Y	L			E	

Key words are: XRAY, GAMMA, IONISE, STERILE, TRACER, CELL

c cells, cancer, space, Sun, particles

Electrostatics – pages 86 and 87

a Second and fourth statements ticked

b spark, air, dangerous, conductor, ground

c

T	H	G	I	L			L
E					P		A
A	D	R	U	M	O		S
T					W		E
I					D		R
N	I		M	A	G	E	
G						R	

Power of the atom – pages 88 and 89

a **i** experiments **ii** kilograms **iii** larger
iv decreasing

b ii, iii

c The fuel used in a nuclear reactor is uranium-oxide; A moderator is used to slow down the neutrons; Control rods are made from either boron or cadmium; In a nuclear reactor the chain reaction is controlled

d energy, temperature, repel, mass

Biotechnology – pages 90 and 91

a Circled conditions: 37 °C, High concentration of reactant, Specific pH for each enzyme

b False, true, true, true

c Circled qualities: eye colour, hair colour, cystic fibrosis, sex, cancer

d Aspirin: Found in the bark of the willow tree; Taxol: Found in the bark of the Pacific Yew tree; Quinine: Extracted from the bark of the cinchona tree; Artemisinin: Extracted from the Chinese tree, Artemsia annua

Animal behaviour 1 – pages 92 and 93

a Learned, instinctive, learned, instinctive, instinctive, learned

b communicating, pheromones, facial, body, humans, aware, emotions

c more, carnivores, energy, groups, numbers, running

d **i** Any organism that feeds on other living organisms and cannot produce its own food
ii The process by which large insoluble molecules are broken down into small, soluble ones
iii An organism that feeds on other organisms
iv Many carnivores live quite solitary lives
v An organism that feeds on plants
vi Many vertebrate herbivores feed in a herd

Animal behaviour 2 – pages 94 and 95

a sexual, courting, dance, song, fight, pheromones, fittest

b i Offspring usually stay with the mother for longer

ii Increased by the length of time spent by the parents rearing the young

iii The length of time the mother is pregnant

iv Usually less time invested in each one

c Food: Cow; Clothing: Mink; Hunting: Fox hound; Companionship: Domestic cat; Entertainment: Circus animals and racehorse

d Answers will vary

Chemical detection – pages 96 and 97

a H_2O: OH^- ion solution, H^+ ion solution; NH_3: OH^- ion solution, ammonium salt, CO_2: H^+ ion solution, carbonate; H_2: magnesium, H^+ ion solution

b Orange, Yellow, Green, Lilac, Yellow

c $2OH^-_{(aq)}$, $Ca(OH)_{2(s)}$, $Al(OH)_{3(s)}$, $2OH^-_{(aq)} \rightarrow Fe(OH)_{2(s)}$, $3OH^-_{(aq)} \rightarrow Fe(OH)_{3(s)}$, $Zn^{2+}_{(aq)}$

d Cl^-: dilute nitric acid, silver nitrate; CO_3^{2-}: dilute HCl, limewater; SO_4^{2-}: dilute HCl, barium chloride; SO_3^{2-}: dilute HCl, acidified potassium dichromate(VI); I^-: dilute nitric acid, silver nitrate

Using equations – pages 98 and 99

a 2, 2, 2, 4, 0.25, 20 g, 20 g, 9 g, 34 g, 4.4 g

b 4 g, 1 g, 8 g, 17 g, 88 g, 56 g, 8 g, 106.5 g, 6 dm^3, 72 dm^3, 6 dm^3, 6 dm^3, 48 dm^3, 6 dm^3, 48 dm^3,

c 2 M, 1 M, 4 M, 0.2 M, 0.5 M, 20, 126, 7.3, 70, 196

d 0.5 M, 1 M, 1 M, 2 M, 1 M, 0.8 M, 1 M, 0.005 M

Chemical products 1 – pages 100 and 101

a Water pipes: unreactive; Pigments: coloured compounds; Dyes: coloured pigments; Wiring: good conductor; Heat exchanger: good conductor; Lamp filaments: high melting point; Needles and nibs: hard; Catalytic converter: speeds up reactions; Electrical contacts: unreactive and non-corroding

b Acid, alcohol; acid; ester; acid; acid, alcohol

c Oxidation, reduction, reduction, oxidation, reduction, oxidation

d Negative, positive, negative, positive, negative, positive, negative, positive

Chemical products 2 – pages 102 and 103

a Transition, transition, alkali, both, transition, transition, alkali, alkali, transition

b i alkali, corrosive, dissolves grease

ii soap, dissolves grease

iii made from fibres

iv alkali links wood fibres

v sodium carbonate

vi alkali, washing soda, dissolves grease

c sulphur dioxide, oxygen, vanadium oxide, 450, 2, sulphur trioxide, concentrated, 98%.

d Soap: sodium salt of fatty acid, can form scum; Hydrophobic: repels water molecules; Bleach: in biological detergents; Enzymes: in biological detergents; Hydrophilic: attracts water molecules; Detergent: in biological and non-biological detergents, sodium salt of fatty acid, surfactant; Oil: ester of glycerol; Ca^{2+}: makes hard water

Particles in action – pages 104 and 105

a Second and third statements ticked

b Fundamental – electron, up quark, positron; Not fundamental – proton, neutron

c X-rays, voltage, charge

Medical physics – pages 106 and 107

a False, false, true, false, true

b i, iv, v

ANSWERS TO EXAM-STYLE QUESTIONS

Biology B1a and B1b

1 c
2 c
3 c
4 d
5 b
6 d
7 b
8 b and d

Chemistry C1a and C1b

1 c
2 c
3 c
4 c
5 b
6 d
7 a
8 c
9 d
10 b
11 c
12 d

Physics P1a and P1b

1 c
2 c
3 d
4 b
5 c
6 a
7 c
8 a
9 d
10 a
11 a
12 d

Additional Biology

1 b
2 d
3 b
4 b, d, c, a
5 b
6 b
7 c
8 d

Additional Chemistry

1 a
2 a
3 b
4 b
5 b
6 d
7 d
8 b
9 b
10 a
11 c
12 d

Additional Physics

1 b
2 c
3 b
4 c
5 b
6 c
7 c
8 d
9 b
10 a
11 b
12 a

GCSE Biology

1 d
2 a and d
3 b
4 d

GCSE Chemistry

1 c
2 d
3 c
4 c
5 a
6 d

GCSE Physics

1 a
2 a
3 d
4 b
5 d
6 a
7 c
8 b